AF323669

Written by Peter Christie
Creative Direction by Leslie Miles, gordongroup marketing + communications
Design by Kelly Read-Lyon, gordongroup marketing + communications

Canadian Foundation for Climate and Atmospheric Sciences
The Sky's the Limit: Ten Years of Achievements
by the Canadian Foundation for Climate and Atmospheric Sciences

ISBN 978-0-9689270-2-1

Printed in Canada

Published in Canada by
Canadian Foundation for Climate and Atmospheric Sciences
350 Sparks Street, Suite 901
Ottawa, ON K1R 7S8
www.cfcas.org

Canadian Foundation for
Climate and Atmospheric Sciences

THE SKY'S THE LIMIT

Acknowledgements

CFCAS acknowledges the support of the Government of
Canada for targeted research on weather and climate through
the Foundation, and of the scientific development, education
and training this has made possible.

It warmly thanks Peter Christie for writing the text of this
book and so eloquently conveying the importance and benefits
of the work and of the partnerships which CFCAS has had the
opportunity to support.

But none of this would have been possible without the exper-
tise, skills, intellectual rigour and unflagging commitment of a
host of scientists, researchers, technicians, students, administra-
tors and policy-makers, from all sectors and from across the
country. The Foundation has been privileged to work with
the community in the advancement and application of new
knowledge on our weather, climate and oceans. This book
is a celebration of their accomplishments.

The Canadian Foundation for
Climate and Atmospheric Sciences
(CFCAS) enhances Canada's
scientific capacity by funding the
generation and dissemination
of knowledge in climate and
atmospheric science areas of national
importance and policy relevance,
through focused support for excellent
university-based research.

CONTENTS

THE SEARCH FOR ANSWERS

TEN YEARS AGO, the creation of the Canadian Foundation for Climate and Atmospheric Sciences (CFCAS) was announced in a federal budget speech that sounded at times like a rallying cry:

"Make no mistake," said the finance minister. "If we are to successfully tackle climate change, if we are to cut costs and boost productivity, if we are to transform ourselves into the world leader in the field of clean energy, then we have to employ every bit of the skill and knowledge we possess."

The inevitable consequence of change is a growing number of questions—how? why? what does it mean?—and the time had come to ensure Canada was a leader in the hunt for answers.

CFCAS has been on the frontier of this search for the past decade, working "to employ every bit of the skill and knowledge we possess." As Canada's premier funding agency for university-based climate and atmospheric research, CFCAS has helped scientists across the country ask and answer the questions that matter to Canadians—questions about the weather, the climate, and the air that affect every one of us.

IN THE SKY, NOTHING IS SIMPLE: weather is a complicated daily brew of temperature, wind, and humidity, while climate is the long-term average weather of a place—part of an intricate globe-spanning system that involves the churning of oceans, the features of landscapes, the chemistry of air, and even the peculiarities of our orbit.

The weather always changes, but these days, the climate is changing fast, too. These changes are complicating things—affecting everything from the air we breathe and the crops we grow to our economic opportunities and challenges.

CFCAS has been tackling this daunting complexity through its support of unprecedented research collaborations, bringing together leading experts from among 436 Canadian scientists in dozens of different fields—from meteorology to marine chemistry, from physics to farm science.

Through research grants totalling more than $117 million at 37 Canadian universities and more than $150 million in cash and in-kind support from partner organizations, CFCAS has funded two dozen Canada-wide research networks and more than 170 major scientific projects. It is wide-ranging work that has revealed important insights into everything from storms and droughts to air pollution, weather forecasts, and climate change.

In just 10 years, the questions answered by CFCAS-supported science have made a difference to the lives of Canadians: they are helping northerners respond to their warming landscape; they are helping prairie farmers prepare for dry spells and avoid the worst effects of billion-dollar droughts; they have equipped people in the Maritimes to anticipate Atlantic storms that often batter coastal communities.

A third of CFCAS research support—$30 million—has been devoted to the North, helping to reveal the mysteries of shrinking arctic sea ice, melting permafrost, arctic pollution, and the melting of Canada's northern and western glaciers. Almost one-fifth—$20 million—has supported work that models distinctly urban weather and climate, establishing clear links between air quality, atmospheric chemistry, and local weather.

JUST AS CANADA has seen greater climate and weather changes than many other nations around the world, some of the most pressing questions reside here, too.

As the planet's second largest country (by area) with seven percent of its forests and long stretches of coastline, Canada is steward to great natural cycles and stores of carbon dioxide and other greenhouse gases. Meanwhile, our Far North has a special influence over the world's oceans and its climate. How these things are managed will have a lasting impact not only here but in other parts of the planet as well.

While 10 years of CFCAS-supported research has equipped scientists and policy makers with the information they need to better manage these resources, the future is far from certain. Our changing climate means we need science to help us adapt but also to help us take advantage of the new opportunities for innovation and leadership that inevitably accompany change. That's why half of all CFCAS funding has helped support and train more than 1,500 graduate students, post-graduate students, post-doctoral fellows, technicians and others to create the next generation of world-class climate scientists.

"GIVEN THE IMPORTANCE of natural resources to our country and because of the severity of our climate, leadership … is not a matter of choice for Canada," said the minister before announcing the federal funds to launch CFCAS in 2000.

Ten years later, the government's statement is no less true. While this book is a look back at the past decade of CFCAS leadership and accomplishments in weather, climate, and atmospheric science in Canada, it is also a look forward.

The story told here is of many climate puzzles solved; but more arise every day. Recognizing these questions and continuing to ask them is vital to our future. More than ever, we need the work of CFCAS to keep Canada's best climate scientists working together to find answers.

Changing Canada

Annual mean temperatures and precipitation increased in most parts of Canada during the 20th century.

Canada is expected to **warm more than the global average** in the coming century with the greatest warming in the Arctic.

Before the turn of the next century, a medium climate scenario predicts **winter in the high Arctic will be 7° Celsius warmer**. Winter will be 3° to 4° Celsius warmer in Ontario and southern Quebec and about 2.5° Celsius warmer in coastal British Columbia.

Summer warming of between 2.5° and 3.5° Celsius is expected over most of Canada. (The Prairies and the interior of British Columbia will be warmer than this average and the Arctic coast will be cooler.)

TEN YEARS OF CFCAS

December 11, 1990
The federal government's "Green Plan" includes funds for what is now known as the Meteorological Service of Canada, to establish a new Canadian Climate Research Network to coordinate university and government climate science across the country.

1993
The Canadian Climate Research Network begins operations.

April 29, 1998
Canada is among the first of more than 180 countries to sign the Kyoto Protocol. Four years later, Parliament ratifies the agreement, committing Canada to reduce greenhouse gas emissions by six percent from 1990 levels.

February 28, 2000
The federal budget earmarks $60 million over six years to create the Canadian Foundation for Climate and Atmospheric Sciences (CFCAS). The goal of the autonomous foundation is to "enhance Canada's scientific capacity by funding the generation and dissemination of knowledge in areas of national importance and policy relevance, through focused support for excellent university-based research in climate and atmospheric sciences."

2001
The Canadian Climate Research Network—effectively subsumed by CFCAS—officially closes its doors. CFCAS attains charitable status.

2003
CFCAS receives a further $50 million from the federal government, boosting the foundation's budget to $110 million and extending its life to March, 2011.

July 2006
The budget for CFCAS research grants is fully committed. (Grants funded after 2006 are paid for through earned interest on CFCAS investments.)

2007
The Fourth Assessment Report of the Intergovernmental Panel on Climate Change (IPCC) is released.

September 2009
The federal government grants a one-year, no-cost extension to CFCAS from March 2011 to March 2012. No new research funds are provided, but the extension allows CFCAS networks and projects to continue their work until March 31, 2011.

June 2, 2010
CFCAS holds a session at the joint Ottawa meeting of the Canadian Meteorological and Oceanographic Society (CMOS) and the Canadian Geophysical Union (CGU). The session is entitled *CFCAS Achievements – The First Decade*, and highlights the impacts of CFCAS-funded research on climate, weather operations, policy development and decision-making.

CANADA'S WORLD-LEADING CLIMATE SCIENTISTS

"Canada is really at the forefront of climate and atmospheric science in the world," says Dawn Conway. "Canadian researchers have become leaders in their fields compared to their peers from elsewhere."

As executive director of CFCAS since its early days a decade ago, Ms. Conway has had a front-row seat watching Canadian researchers take centre stage in international efforts to understand weather and climate and to apply scientific information to Canadian needs. Few other countries have been as successful at ensuring their best minds are tackling the world's most pressing environmental problem.

"It means Canada often finds itself involved in visioning exercises at the level of the World Climate Research Programme or the International Council for Science or other initiatives," Ms. Conway says. "It also means working alongside other world leaders in almost every area of this important research."

Many CFCAS-supported Canadian scientists, for instance, participated on the Intergovernmental Panel on Climate Change, which was awarded the 2007 Nobel Peace Prize jointly with former U.S. vice-president Al Gore. Many CFCAS-funded researchers have also been honoured with other international and national accolades.

For example, Richard Peltier who heads the Polar Climate Stability Network (see Chapter 4) was awarded the 2010 Bower Award and Prize for Achievement in Science (the leading U.S. prize in science includes a $250,000 purse). Meanwhile, the chair of the CFCAS Board of Trustees, Gordon McBean, was inducted as a member of the Order of Canada in 2010. Several other CFCAS network and project investigators have been recognized through other science awards, and many hold prestigious Canada Research Chair positions at their respective universities.

In some cases, CFCAS-funded research programs helped attract these leading scientists to their Canadian universities in the first place. In other cases—and perhaps most importantly—the best upcoming climate researchers got their start and were mentored while working on CFCAS-supported networks and projects.

During the past 10 years, more than $58 million of CFCAS research support has contributed to the training of almost 1,500 undergraduate and graduate students, post-doctoral fellows, research associates, and technicians— the highly qualified people Canada needs to respond to the opportunities and challenges of our changing climate.

"One of the most important things we've done as a Foundation has been to help train the next generation of climate and atmospheric scientists," says Ms. Conway.

More than 50 cents of every CFCAS research dollar has gone toward training or supporting almost 1,500 undergraduate and graduate students, post-doctoral fellows, research associates, and technicians. According to a 2009 review of CFCAS contributions to Canada's highly qualified people (HQP), CFCAS support for preparing Canada's future weather and climate researchers was expected to top $58 million by 2010.

Of 700 people trained through 87 completed CFCAS initiatives during the Foundation's first four years, a quarter went on to take research jobs with the federal government (Environment Canada, Fisheries and Oceans Canada, Natural Resources Canada, etc.), with industry, or with universities. More than a third continued their education, pursuing further university degrees.

In 2007, the federal government released its science and technology strategy recognizing that "talented, skilled, creative people are the most critical element of a successful national economy over the long term." Today, this call to create a national "knowledge advantage" is especially important for Canada, as our fast-changing climate creates new and often unexpected opportunities and challenges.

For the last decade, CFCAS has answered this nationally important challenge by attracting, supporting, and fostering climate and atmospheric scientists who are second to none around the world.

1 THE
STORMS

No one knows better than the people of Pangnirtung, Nunavut that Canadian storms are not what they used to be.

On June 8, 2008, black clouds and high winds rolled over the Baffin Island town, unleashing a downpour as fierce as it was strange. The rain and the unseasonably mild temperatures melted hilltop snow so quickly it filled the Duvall River, sending a wall of water surging into the hamlet.

A 10-metre channel quickly carved itself through town, deep into the permafrost and down to the bedrock. A road collapsed. Huge sinkholes appeared. Two bridges teetered as the torrent eroded their foundations then they toppled altogether.

"It is something that is new to us," said Ed Zebedee, director of Protection Services for the Government of Nunavut. "We're actually looking very carefully at it. We're talking to our experts."

The unusual storm took everyone by surprise, battering the little town with an estimated $5 million in damage. Many people were without water or sewage for days. Officials declared a local state of emergency, and hunters used freighter canoes to ferry bewildered residents through the churning brown water.

WELCOME TO THE NEW NORMAL, says veteran storm-watcher and University of Manitoba professor John Hanesiak. "What we're finding is, generally, an increase in more intense storms in the Arctic."

Researchers call it extreme weather—bizarre flash floods, sudden rain storms, blizzards, tornadoes, hail, and heat waves. Although hazardous weather is nothing new to Canadians,

▲
Researchers with STAR and members of the community of Iqaluit collaborate to make extreme weather more manageable in the North.

extreme weather is often devastating because it is out of the ordinary and unexpected. And, thanks to our changing climate, it's becoming more severe and more common, not just in the Far North, but across Canada and in other parts of the world.

For many, extreme weather is like the vanguard of global warming: unlike slower climate changes whose effects may not be felt for years, dangerously strange weather is challenging Canadians now—and hitting them where they live.

An 11-year-old boy, for instance, was killed at a nature camp when tornadoes swept through Durham, Ontario in 2009—an event that startled many and led Ontario Premier Dalton McGuinty to call for better weather warnings. A wild wind

Major weather-related catastrophes around the world caused more than US$750 billion in total losses between 2000 and 2008, according to insurance giant Munich Re.

that hit Vancouver in 2006 felled almost half the forest in the city's world-famous Stanley Park. Severe Manitoba floods in 2009 at one point covered more than 700 square kilometres.

Brought about by climate-system-related shifts in storm tracks and hydrological cycles, extreme weather is also costing us more. According to global insurance giant Munich Re, the number of major weather-related catastrophes worldwide has tripled since the early 1980s. Weather disasters caused more than US$750 billion in losses between 2000 and 2008 alone.

The future, according to the Intergovernmental Panel on Climate Change, likely promises more costs—and more questions.

"There are a lot of unknowns," says Dr. Hanesiak.

CFCAS HAS BEEN FUNDING SCIENCE to get to the bottom of these extreme weather unknowns for the better part of a decade. CFCAS-supported research networks and projects have steadily improved our understanding of what's behind severe storms and other out-of-the-ordinary weather through research into events as varied as blizzards in the Arctic, storms along our marine coasts, droughts in the Prairies, and heat waves in Canada's major cities.

These CFCAS-funded initiatives have described the processes behind severe storms and familiarized a whole new generation of scientists with weather extremes. Most importantly, they've helped develop tools that forecasters and modellers can use to warn Canadians before wild, hazardous weather suddenly puts us at risk.

The CFCAS-funded **Storm Studies in the Arctic** (STAR) network, for example, is a five-year (2005-2010), $3 million scientific network headed by Dr. Hanesiak and his University of Manitoba colleague Ron Stewart.

The network—involving dozens of scientists and graduate students from universities and federal and territorial governments across Canada—combines observations of snow, wind, sea ice, and land topography with other data on storms gathered from weather instruments and research flights around Pangnirtung and the Nunavut capital of Iqaluit in late 2007 and early 2008.

The result, says Hanesiak, has been an unprecedented understanding of the processes and systems at work behind dangerous arctic weather. Careful measurements of the shape and size of snow crystals, for example, has revealed what makes some

Wired for ice

Beginning on January 4, 1998, a six-day ice storm pelted much of eastern Canada, leaving some cities in the region glazed with the icy equivalent of 100 millimeters of frozen rain. The crushing weight of ice toppled some 1,300 transmission towers, 27,000 utility poles, and enough cable and wire to stretch around the world three times. More than a million Quebec residents and about 100,000 in Ontario were without power for days as a result.

While no one could have predicted the extent of the ice storm, understanding how and why ice forms on power lines and towers could equip utility companies to face another icy onslaught in the future. That's why a CFCAS-funded project at the Icing Research Group at the University of Alberta (along with researchers with the Icing Precipitation Simulation Laboratory at the Université du Québec à Chicoutimi) spent three years working to develop and improve freezing rain accretion models for poles, lines and towers. Led by University of Alberta professor Ed Lozowski, the project helped improve predictions of how and when ice forms and accumulates on the frame structures and wires—information power and transmission industries can use when designing towers that can take the icy weight.

blizzards blinding. The work also helps explain what's behind other hazards, such as severe wind chill. It's invaluable information for meteorologists to help improve the weather warnings that many northerners depend upon.

Using community surveys, STAR scientists are also able to ensure their research addresses climate questions that people in the Arctic considered most pressing—such as regional differences in severe weather systems and the local impact of blizzards.

BUT CFCAS-SUPPORTED EFFORTS to understand extreme weather are not limited to the North. The groundbreaking work of two earlier five-year research networks helped improve the forecasting of severe weather elsewhere in Canada.

A nearly-$3-million network, **Improving Quantitative Precipitation Forecasts of Extreme Weather**, developed improved tools for predicting when, where, and how much rain or other precipitation to expect with oddball storms and extreme weather. The **Enhanced Short Term Forecasting of Extreme Weather** network, meanwhile, directed its many researchers and almost $2.5 million in CFCAS support to improving short-term forecasting accuracy in cases of sudden severe weather.

Smaller CFCAS-funded projects have supported extreme-weather scientists with interests as varied as forecasting squalls and storms in the Great Lakes, understanding how ice forms on airplane wings or transmission towers, simulating prairie hail storms and tornadoes, or modelling how snow drifts and blows.

Recent Extreme Weather Disasters in Canada

Hurricane Igor tore into Newfoundland in September 2010. A deluge of rain washed out roads and bridges, severed the Trans-Canada highway and flooded houses. Gale-force winds damaged roofs and downed trees and power lines.

In 2009, **tornadoes in Southern Ontario** caused the death of an 11-year-old boy and led Premier Dalton McGuinty to call for better extreme-weather warning systems.

In 2006, a severe windstorm ripped through **Vancouver's world-famous Stanley Park**, flattening 3,000 trees and causing $9 million in damage.

A tornado cluster caused about **$100 million in damage to Peterborough**, Ontario and the surrounding area in 2005.

A severe downpour in Toronto in 2005 led to the largest insurance payment to homeowners in Ontario's history. On a national scale, it was second only to the ice storm of 1998.

On September 29, 2003 **Hurricane Juan** slammed into Nova Scotia, causing $300 million in damage across the centre of the province and into Prince Edward Island.

In British Columbia, 300 homes were destroyed and 45,000 people evacuated when the hot, dry summer of **2003 fuelled huge forest fires.**

A once-in-a-century violent rainstorm flooded Peterborough, Ontario twice in two years (2002 and 2004).

In 2001 and 2002, **prairie drought cost $3.6 billion in farm losses** and left more than 41,000 people out of work, according to the Saskatchewan Research Council.

A January 2000 storm surge washed over parts of Prince Edward Island and other communities, flooding homes and basements and setting **a new record water level in Charlottetown**.

In 1997, Manitoba suffered its second-worst flood ever recorded and in 2009 suffered its third-worst flood—at one point **drowning more than 700 square kilometres of land**.

The 1998 ice storm in eastern Canada was blamed for the deaths of 28 people and more than 900 injuries. The cost of the disaster exceeded $5 billion and resulted in the largest insurance payout in Canadian history.

Flooding in the Saguenay-Lac-Saint-Jean region of Quebec in 1996 destroyed more than 1,000 homes and required 16,000 people to be evacuated.

2 THE
DROUGHTS

B*y July 2002, drought had Alberta rancher Jay Fenton (lower right) against the ropes.*

The beleaguered farmer realized that the parched pastureland on his century-old family farm two hours southeast of Edmonton could no longer feed his 1,200 cattle. He was left with no choice: all but fewer than 200 of the generations-old herd had to be sold at auction.

"It is a family heirloom … You hand it down from generation to generation," he told reporters at the time. "This is going to be a challenge that we've never, ever experienced."

Mr. Fenton and his family were not alone on unfamiliar, heartbreaking ground. By the time of the cattle auction, farms throughout the Canadian Prairies were reeling from almost three years of drought. Another two years of severe dry weather were yet to follow.

ALTHOUGH 40 DROUGHTS HAVE HIT western Canada in the past two centuries, the five-year dry spell between 1999 and 2004 has been called the most expensive natural disaster in Canadian history. By 2001 and 2002, water shortages were being felt right across the country, and some prairie regions were suffering from a drought more severe than anything they had seen in a century.

Ranchers like Mr. Fenton, along with wheat farmers across Saskatchewan and Alberta, struggled to keep their operations alive. Hay and cattle feed were shipped from as far east as Prince Edward Island.

"There's no question it was a terrible time for many people," says University of Manitoba scientist and geographer Ron Stewart.

▲ By August 2002, farmer Brad Lockhart and his wife, Sandy, realized their crops were lost for the second year in a row. The third-generation farmer from Youngstown, Alberta, was hard-hit by the five-year drought between 1999 and 2004.

▼ Drought left Alberta rancher Jay Fenton with a diminishing stock of feed and no choice but to sell most of his cattle.

What's worse, perhaps, is its portent for the future. According to the Intergovernmental Panel on Climate Change, many Atmosphere-Ocean General Circulation Models show regions in northern middle latitudes—including the Canadian Prairies—becoming increasingly dry in the summers (punctuated by rare but intense rains) and more likely to be hit by drought. Some fear climate changes may slowly turn the interior plains of North America into arid scrubland that's potentially too parched for cattle or crops.

That's where CFCAS comes in.

THANKS TO CFCAS-SUPPORTED research networks and projects, scientists now have a better understanding of what's behind the climate shifts and precipitation patterns—including their interactions with soils, crops, and the atmosphere—that lead to prolonged dry spells. The research has uncovered key facts about how and why Canadian droughts happen and when to expect them—significant discoveries in what is a surprisingly little-studied and poorly understood phenomenon.

The **Drought Research Initiative** (DRI) is a $3.2 million CFCAS-supported network that examined in detail the extraordinary drought of 1999-2004, a crisis that has often been compared to the Depression-era drought of the "Dirty Thirties." The research involves 15 university investigators and scores of federal and provincial collaborators representing a variety of disciplines from across Canada. Their work has created an unprecedented in-depth picture of the forces and factors involved in that devastating dry period.

▲ Scientists predict that drier weather on the Prairies will nevertheless be punctuated by occasional severe rainstorms. During several days of torrential June rain in 2010, floods washed across parts of southern Manitoba, Saskatchewan (*above*), and Alberta, inundating homes, washing out bridges, and closing the Trans-Canada Highway.

◄ Equine rancher John Ruzicka said feed costs for his Percheron horses would top $100,000 after the pasture land at his farm in Killam, Alberta withered during a 2002 drought.

The Prairies' parched past

Are the Prairies on their way back to the future? Some climate watchers fear recent climate changes may signal a return to periods in the past when western Canada was far more desert-like than it is today. Six millennia ago, for instance, sand dunes drifted across parts of southern Saskatchewan, prairie grasslands grew farther north and east, and many prairie lake basins held no water at all.

Understanding these earlier climates could be critical to anticipating what might happen next. CFCAS supported Brian Cumming and other researchers at Queen's University to study proxy records of past droughts—including their severity and duration—going back 6,000 years. The team pulled cores of lake-bottom sediment to look at the trapped shell remains of single-celled organisms called diatoms. Because different diatom species represent a variety of climate-revealing conditions (e.g., lake salinity), the researchers were able to reconstruct climate changes at several time scales.

The scientists found that the relatively moist prairie climate of today may be as young as 1,700 years old—the last of as many as three significant climate shifts in 6,000 years. They also found climate influences have altered the eastern and western regions of the Prairies somewhat differently through history.

The CFCAS-supported Drought Research Initiative's detailed portrait of the worst prairie drought in a generation is helping forecasters understand when a drought will strike and how long it might last. A January 2010 workshop gave the federal and provincial governments a firsthand look at the predictive powers of this groundbreaking research that could save millions in farm losses.

"The more we understand one drought and why the [forecasting] models may have failed, the more we can have confidence in better prediction for the future," says Dr. Stewart, who is co-leader of DRI with John Pomeroy of the University of Saskatchewan.

By comparing the 1999-2004 drought to longer-term data and to studies of other dry periods, DRI is revealing elements behind the mechanics of drought that forecasters can use to better predict rain shortfalls and to provide farmers with a more reliable weather outlook. The work has also helped to fill a significant gap in international drought science concerning the role of land-use changes, agricultural and tilling practices, and the mechanics of drought. In 2011, a special issue of Canada's scientific journal *Atmosphere-Ocean* is expected to feature much of this ground-breaking DRI research.

SOME CFCAS NETWORKS HAVE EXPLORED the mysteries of drought in other contexts. Among these, the **Western Canadian Cryospheric Network** (WC2N) spent five years documenting—for the first time—the extent of glaciers in the mountain ranges of British Columbia and Alberta and how changes to this mountaintop ice will affect water and energy cycles. The glaciers are important to hydro power on the West Coast (see Chapter 5), but they are also a significant source of melt-water to rivers supplying Alberta farmlands, especially during the dry periods of late summer.

Similarly, the CFCAS-supported network known as **Improved Processes and Parameterisation for Prediction in Cold Regions** (IP3) has helped to describe and better understand the interactions of water cycles and the atmosphere in cold regions such as the Rocky Mountains and northern territories—interactions that also affect available water to downstream farms, industries, and towns (see Chapters 4 and 5).

Other smaller CFCAS-supported projects targeted features of prairie drought and water shortages at both global and regional scales. El Niño climate cycles, for instance, are linked to recurring drought conditions in the North American Midwest. That's why CFCAS-funded science has been exploring ways to better predict El Niños and to understand how these climate events influence changes in stream flows throughout western Canada.

CFCAS projects have also worked to better project the availability of water, to understand the role of soil moisture, to better predict hail, heavy rain, and other severe weather, and even to gauge the amount of water vapour in the air using Global Positioning Satellite (GPS) signals.

The drought of the century

The years 1999 to 2004 were marked by successive **below-average rain and snow fall** in many regions in the Prairies.

Several areas experienced the most severe drought conditions **in more than a century**. In 2001, for example, Saskatoon was a third drier than in any year since 1891.

Farm production plummeted by about $3.6 billion in 2001-2002, contributing to a $5.8 billion drop in Canada's Gross Domestic Product (GDP).

Most of the losses occurred in the 2002 season, with $2 billion in farm losses contributing to a **$3.6 billion fall in GDP**.

In 2001-2002, more than **41,000 people** found themselves out of work on the Prairies as a result of the drought.

Net farm income was negative or zero for several provinces for the first time in 25 years. (A negative net farm income occurred in P.E.I. for 2001, in Saskatchewan for 2002, and a zero net farm income was reported for Alberta in 2002.)

Cattle and other livestock herds had to be **sold off or slaughtered**, especially in Alberta.

Impacts that could take **a decade or longer to recover** from included soil and other damage by wind erosion, and deterioration of grasslands.

By 2001, drought effects were felt across the country, with water-use restrictions in place in British Columbia, record low lake levels in the Great Lakes-St. Lawrence River system, and 50-100 percent crop losses on Prince Edward Island.

CFCAS-funded researchers with the IP3 network study ice, water and ▶ weather systems in the Rocky Mountains and the western Arctic.

3 THE CITIES

B

rianna Nudds learned how frighteningly different city weather can be while she was driving through Hamilton, Ontario on July 26, 2009.

Ms. Nudds, her sister, and her sister's two young children met a severe thunderstorm as they entered the city, and very soon their compact car ploughed headlong into a flooding underpass. Rushing rainwater swept the vehicle to the bottom of the asphalt gully, and the women and children narrowly escaped by climbing through the car windows.

"It was really scary," Ms. Nudds told reporters. "I was terrified for the lives of my family."

The colossal summer storm had effectively stalled over the heavily developed western end of Lake Ontario, and the city of Hamilton, in particular, was being pummelled by rain and lightning.

The tempest knocked out power for thousands and a growing deluge flooded more than 7,000 homes and businesses with insurance losses of between $200 and $300 million. Radar estimates suggest 109 millimetres of rain fell in just two hours—one of the most intense short-duration rainfalls on record in Canada, according to Environment Canada.

"CITY WEATHER BEHAVES DIFFERENTLY," says University of Western Ontario climate scientist James Voogt. "Just about every characteristic that we use to define weather or climate is altered by urban surfaces."

Rain, storms, wind, and heat are all influenced by the decidedly different weather impacts of asphalt, concrete, and skyscrapers. So is the quality of the air we breathe. Cities can slow a weather front or encourage cloud growth. They can help spur severe rain and

◀ By improving the ability of meteorologists and forecasters to predict unique city weather, such as smog and urban heat islands, the CFCAS-funded EPiCC network has created tools to help health workers and municipal officials prevent heat stroke and other weather-related illnesses from affecting millions of city-dwelling Canadians.

thunderstorms, or they can generate a plume of warmer, more polluted, and often drier air that drifts over landscapes downwind.

City weather is also closely connected to people. Four out of five Canadians live in urban areas, and weather effects on transportation and businesses can have serious economic and social impacts. Meanwhile, air pollution in cities affects the health of millions every year, and the activities of city dwellers affect the climate as tonnes of urban greenhouse gases drift skyward.

Yet, despite all this, urban differences are rarely or poorly accounted for in Canadian weather and climate models. The shortfall, meanwhile, means a significant gap in the weather forecasts that millions of people depend on.

That's why CFCAS-funded researchers have been taking a hard look at urban environments and their relationship to the sky.

CFCAS-SUPPORTED RESEARCH has uncovered the complex role of cityscapes and energy cycles in influencing local and regional weather, climate, and air quality. Thanks to CFCAS-funded science, governments and forecasters have the data they need to help make climate models more accurate and to put urban environments on the weather forecasting map.

The **Environmental Prediction in Canadian Cities** (EPiCC) network, for example, is a four-year, $1.4 million research initiative examining how pavement, buildings, and urban green spaces interact with the air and the atmosphere over Vancouver and Montreal—two very different Canadian cities with equally distinctive climates.

Dr. Voogt is principal investigator for the network, which involves about 20 scientists from three universities and federal and provincial agencies. The study includes the efforts of more than half-a-dozen graduate students and post-doctoral researchers.

Their work uses unprecedented amounts of field data—from weather towers, balloons, specialized lasers, and other instruments—to connect temperature, humidity, and wind measurements

Our share of bad air

The Canada-U.S. Air Quality Agreement became an international environmental success story not long after it was first signed in 1991. The Agreement recognized that air pollution—much of it from urban areas—does not respect national boundaries; pollutants released on one side of the border can easily blow many miles to the other. After being celebrated for reducing acid rain in the 1990s, the Agreement was expanded in 2000 to help cut cross-border smog.

The only problem was figuring out whose bad air was whose.

The CFCAS-funded research of University of Toronto professor Greg Evans has helped arrive at an answer. Using high-resolution measurements, statistical models and information about the composition and origin of airborne particles, Evans and his researchers found a way to distinguish local Canadian pollution from the particles that drifted from the United States during episodes of bad air in Canadian cities.

The results have been important to the scientists and policy makers who continue to monitor and develop the bi-national air quality agreement through biannual report cards. The research is also helping city dwellers breathe a little easier—on both sides of the border.

(*Top*) A CFCAS-funded team working with the Meteorological Service of Canada developed an easy-to-use system for forecasting snow and rain. The system, known by its acronym MAPLE, allows city managers to dispatch snow-clearing equipment before a snow storm arrives or to drain wastewater reservoirs before heavy rain causes an overflow. A similar CFCAS-supported technology, HAILCAST, warns farmers and others when crop-crushing hail is on the way.

(*Bottom*) Doppler Weather Station in Montreal.

A heat wave that scorched Toronto in July 2010 sent Ty Lewis, 8, (*right*) and Ryan Chase, 7, to cool off in the fountains of Dundas Square.

to observations of solar energy, heat exchange, evaporation, and carbon dioxide cycling. The researchers then use this complex suite of information to modify and test a sophisticated forecasting model first developed in France.

It sounds complicated, but EPiCC research is nonetheless revealing a little-known world of weather, air quality, and energy that has a profound effect on Canadian cities—a research breakthrough that will help forecasters and scientists predict city heat waves, sudden urban downpours, changes in air quality, and even how to respond to pollution, including toxic spills or leaks.

"This is something that represents the state of the art," says Dr. Voogt.

WHILE LOCAL CLIMATE IS IMPORTANT to the 25 million Canadians who live in or near cities, so is the quality of the air they breathe. That's why CFCAS-supported networks and projects have also been working to understand the processes that affect smog (airborne particles and ground-level ozone) and its interaction with weather and climate processes. The **Multiscale Air Quality Modelling network**, for example, was a five-year, $4–million effort that used data and computers to successfully model the behaviour of atmospheric aerosols (see Chapter 8). The work is being used to predict changes in air quality at a global scale but also for regions and cities. The network's model is helping scientists and policy makers with the Meteorological Service of Canada and provincial ministries of the environment decide how to respond when air pollution becomes a serious threat to the health of Canadians living in cities.

Meanwhile, CFCAS-funded research also contributed to the Environment Canada-led **Pacific 2001** project to better understand how airborne particles behave over the Fraser River Valley in British Columbia (see Chapter 8).

Smaller CFCAS-supported projects have studied air quality through research questions as varied as how aerosol particles affect clouds, how airborne soot transports toxic chemicals, how harmful airborne chemicals and particles form and grow, how laser radar can reveal patterns of pollution mixing in the air, how pollutants modify the urban climate, and even how forests near cities act as filters of persistent organic pollutants (see Chapter 6 sidebar).

Cities feel the heat

Forty-eight cities in Canada have more than 100,000 residents. **Toronto and Montreal are the country's largest**, with 2.5 and 1.6 million people respectively.

Cities are generally warmer and more likely to be affected by smog than non-urban environments.

In a warmer future climate, there will be an increased risk of more intense, more frequent and **longer-lasting heat waves**.

A phenomenon known as **the urban heat island** means cities can be up to 10° Celsius hotter than the surrounding countryside in some circumstances.

Health Canada estimates that **short-term exposure to air pollution** contributes to 1,800 premature deaths in the country every year. An additional 4,200 Canadians die prematurely each year due to the long-term effects of exposure to air pollution.

4 THE NORTH

J

ust a year after the small arctic community of Salluit in northern Quebec celebrated the opening of its fire station, Paul Okituk and other residents watched in amazement as the newly built hall sank into the ground.

"There is permafrost beneath us, and it's changing," said Mr. Okituk.

With buckled roads, cracked foundations, and sunken buildings, Salluit is just one of many towns and villages throughout the North feeling the effects of melting permafrost. Since the 1980s, temperatures at the top of the Arctic's once-permanently frozen ground have increased by up to 3° Celsius, and the maximum area covered by seasonally frozen earth has decreased by about seven percent throughout the Northern Hemisphere since 1900.

Salluit—which only recently decided against plans to relocate the entire village—is locating any new buildings and housing kilometres away from the downtown, far from the softening turf. Other arctic areas and towns are seeing roadways used by mining and oil industry trucks sag or become impassable.

The thawing and shifting ground is threatening other developments as well. Some worry about possible buckling or shearing affecting oil and gas pipelines, including the long-planned $16 billion pipeline project along the Mackenzie Valley.

THE ARCTIC IS CHANGING. Our warming climate is transforming the Far North more than any other region on the planet. Over the past 20 years, arctic temperatures have been rising almost twice as fast as the global average.

▲ Work in the Far North by the CFCAS-funded Canadian Network for the Detection of Atmospheric Change (CANDAC) has tracked changes to air quality that provide valuable information to efforts to cut toxic emissions in Canada.

Canada's commitment to the Arctic—to its sovereignty, its development, and its environment—was outlined in the government's Northern Strategy in July 2009. In December 2009, the Senate Standing Committee on Fisheries and Oceans called for the creation of an arctic affairs cabinet committee, chaired by the prime minister. In August 2010, the Canadian government announced that a world-class arctic research station would be built in Cambridge Bay, Nunavut.

The accelerated changes are having a dramatic impact on the lives of people and wildlife—an existence uniquely tied to a landscape of cold and ice. The impact does not stop there: climate shifts and ocean changes in the Far North have a significant influence on weather and climate systems elsewhere around the world.

The melting of sea ice over the Arctic Ocean, for instance, is causing a remarkable case of so-called climate "feedback." Millions of square kilometres of bright white ice capable of reflecting warm sunlight back into space is being lost and replaced by the dark sea that absorbs solar heat. The heat is changing the local climate—a large reason for the amplified warming in the Arctic—but temperatures, weather systems, and oceans everywhere are affected.

Warming permafrost, meanwhile, is another unknown. The northern permafrost region—16 percent of the planet's ground cover—contains half of the world's soil-trapped organic carbon and most of this (88 percent) is shut up in perennially frozen ground. The thawing earth could unlock this great carbon storehouse, releasing carbon dioxide to the atmosphere. It could also release underground methane—an even more powerful greenhouse gas.

The challenges facing this great, austere land are particularly important to Canada. As the second largest polar nation in the world, Canada is steward to vast areas of tundra, polar sea, and thousands of arctic islands. Understanding the changes taking place above the 60th parallel—40 percent of Canada's land mass—is vital not only to helping communities, industries and ecosystems in the Arctic but also to improving how we understand weather and climate changes around the globe.

"People get rather confused by the fact that although the Arctic region is a small fraction of the area of the planet, it has a totally disproportionate impact on global climate in general," says University of Toronto physicist Richard Peltier.

FOR THE PAST 10 YEARS, answering the tough weather and climate questions facing Canada's Far North has been a priority for CFCAS. More than $30 million of CFCAS scientific funding—almost a third of the Foundation's research support—has helped scientists from across the country improve our understanding of subjects as diverse as arctic sea ice, melting permafrost, arctic pollution, polar storms, and climate connections to other parts of the planet.

For instance, the **Polar Climate Stability Network** (PCSN) is a five-year, $5-million research network that uses clues about climate variation from the past to help model changes to temperatures, oceans, sea ice, and arctic land glaciers into the future.

Dr. Peltier heads the network, which includes 14 main investigators and dozens of other researchers from 11 universities and government agencies across Canada. These scientists are exploring the Arctic's climate past and future by studying Arctic Ocean seabed sediment, ice cores, tree rings, and data from weather instruments as well as climate models run on supercomputers—including the most powerful computer in the country at the University of Toronto's SciNet facility.

In 2009, PCSN research was the subject of special issues in two leading climate science journals, the *Canadian Journal of Earth Sciences* and the international *Journal of Climate*.

"If Canada is going to make a contribution to climate science then it's only natural that we focus on the region that has felt the most impacts of the global change process," says Dr. Peltier.

THE **CANADIAN NETWORK for the Detection of Atmospheric Change** (CANDAC; see Chapter 8) is another CFCAS-funded network working in the High Arctic, this time to explore the arctic atmosphere, including the role of ozone and other airborne chemicals on climate and air quality. The network uses satellite data and other measurements collected at a remote northern research station at Eureka on Ellesmere Island (see sidebar: PEARL of wisdom).

▲ The CFCAS-supported Crater Lake Project in Pingualuit in northern Quebec provided a first-time look at how organisms once lived in lakes locked under the ice of glaciers. The project has attracted international interest and revealed secrets of arctic climate history going back more than 100,000 years.

Other CFCAS-supported networks explore interactions of the climate and the landscape on changes to northern and mountain glaciers, and storms and severe arctic weather. A $2.5-million network of scientists from eight universities, governments, and private industry from across Canada known as IP3—short for **Improved Processes and Parameterisation for Prediction in Cold Regions**—is studying glaciers, water, and weather in the western Arctic and the northern Rocky Mountains (see Chapter 5).

Meanwhile, the $3-million **Storm Studies in the Arctic** (STAR) network also has its eyes on the Far North. STAR is helping scientists and others better understand and predict the blizzards and sudden storms that are becoming more common in the climate-changed North (see Chapter 1).

Almost two dozen smaller CFCAS-supported projects have been piecing together other parts of the climate and weather puzzle in Canada's Arctic. Researchers have worked to estimate the extent of arctic permafrost melt during the next century; to measure and model the formation of ice clouds or drifting and blowing snow; to reconstruct past arctic climate variability and sea ice changes; to calculate the climate impact of snow cover on the tundra, creeping vegetation changes near the tree-line, or the iron supply in the Arctic Ocean; and to study carbon exchange and environmental changes in the Arctic at the Tundra Ecosystem Research Station (TERS) at Daring Lake, Northwest Territories.

CFCAS-supported efforts have also helped scientists better understand the ocean and climate impact of Greenland glacier melt-water as part of the international Greenland Flow Distortion Experiment (GFDex). Other projects have examined arctic air quality and ozone, including studies of mercury in the Arctic, research into arctic haze and the impact of bromine on ozone, and other projects to model arctic air quality.

On thin ice

In 2009, scientists from Canada, along with colleagues in Norway, Russia, and the United States, released their report from the Arctic Monitoring and Assessment Programme (AMAP) detailing changes during the recent four years in the Far North.

The land: Permafrost is warming fast and thawing at its margins, allowing plants to grow more vigorously and densely. In Russia, the tree line has climbed further up hills and mountains by about 10 metres every year.

Summer sea ice: The area covered by summer sea ice in 2007 was reduced to about two-thirds the size of the 1979-2000 average. Satellite data also suggests it is much thinner. For the first time in existing records, both the Northwest and Northeast passages were ice-free in summer 2008.

Greenland: The Greenland ice sheet has continued to melt in the past four years with summer temperatures consistently above the long-term average since the mid 1990s. In 2007, the area experiencing melt was 60 percent greater than in 1998.

Warmer waters: In 2007, some ice-free areas were as much as 5° Celsius warmer than the long-term average. An influx of warmer water from the Pacific and Atlantic, combined with the loss of reflective, white sea ice, appears to be the cause.

Soot: Black carbon, or soot, is now adding to the warming of the Arctic by creating a haze which absorbs sunlight and darkens the sun-reflecting white of the snow.

PEARL of wisdom

High on one of Canada's most northerly islands—where mid-winter temperatures average a bone-chilling -38° Celsius and summer normally barely creeps above freezing—intrepid CFCAS-supported researchers have been braving the barren, cold landscape for the sake of a revealing view of our climate and atmosphere that can be seen only from the top of the world.

The CFCAS-funded **Canadian Network for the Detection of Atmospheric Change** (CANDAC) is a $5.7-million project—with additional infrastructure, facility and logistical support from the Canada Foundation for Innovation, the Natural Sciences and Engineering Research Council, and federal departments —located at the Polar Environment Atmospheric Research Laboratory (PEARL) on remote Ellesmere Island, just 1,100 kilometres from the North Pole. The remote northerly station provides scientists with research opportunities that can't be found anywhere else on Earth, such as a unique location to track and validate data from passing satellites being used to monitor atmospheric change.

Headed by Dalhousie University physicist James Drummond along with CANDAC researchers from seven universities and federal and provincial agencies, the network has spent the past five years revealing the mysteries of arctic climate, ozone, and pollution. The work explored how airborne pollution from the south is transported into and through the arctic atmosphere, how mountains and airborne ice crystals affect heat transfer from the sun, and how chemical ozone changes in the atmosphere at three main points: in the deep winter, as the sun begins rising above the horizon again, and in the spring.

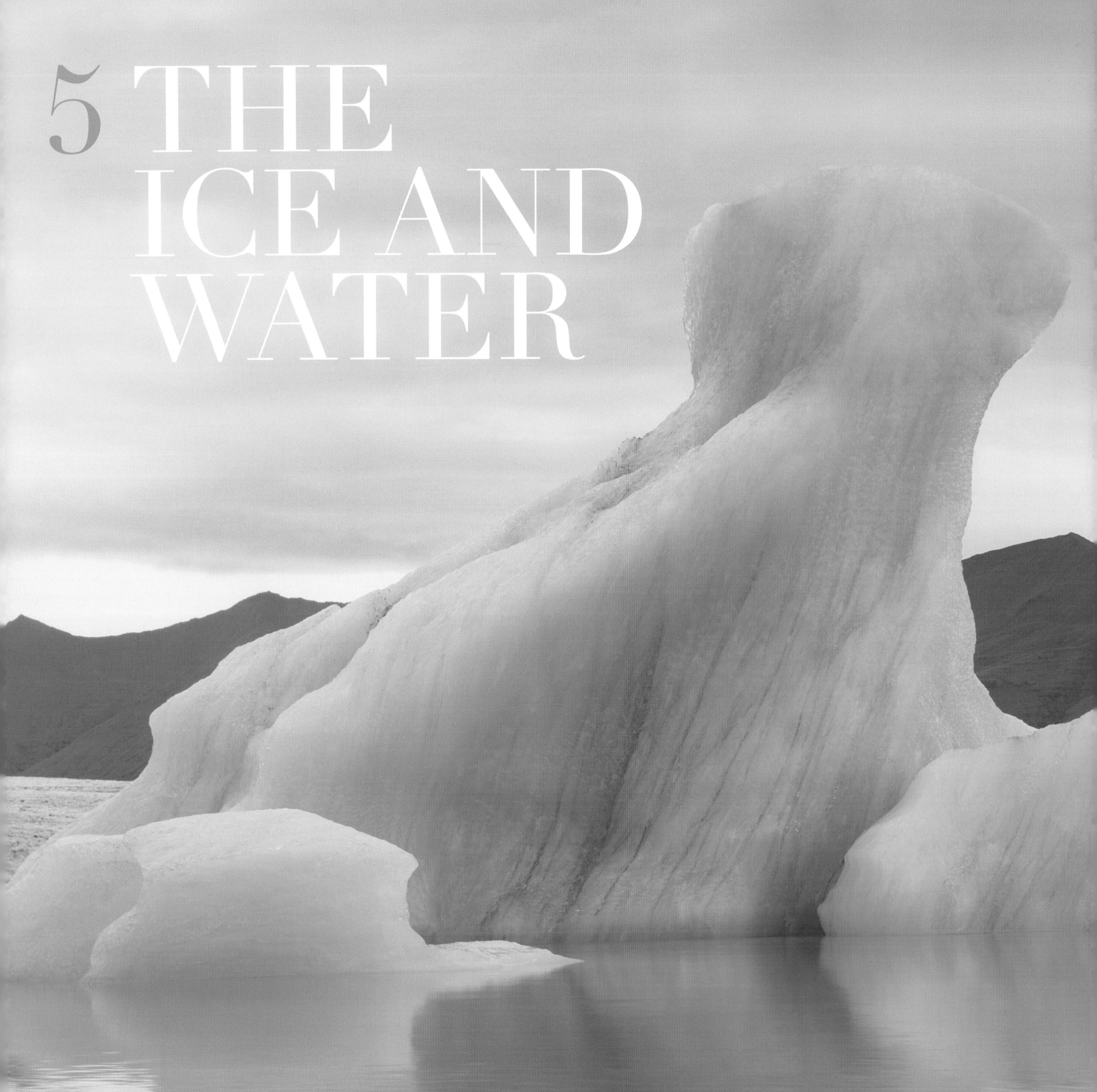
5
THE
ICE AND
WATER

C

anada's chances of honouring its side of a crucial water pact with the United States may be literally melting away.

"There is just a high degree of uncertainty," says Frank Weber, leader of the BC Hydro Runoff Forecasting Team. "Glacier runoff is a concern at this point, but we don't quite know what the impact will be."

In 1961, Canada and the United States signed the Columbia River Treaty to jointly manage the shared, 2,000-kilometre river and its capacity for producing power. The agreement obliges British Columbia to control the river's downstream flow—most of it in the United States—by holding back some thousands of millions of cubic metres of water in three storage dams near the river's mountain origins.

The problem is that decades of melting by British Columbia's mountain glaciers means the water feeding the reservoirs is dwindling. Runoff flowing into the Columbia River system was expected to be about 14 percent below normal for the first eight months of 2010—the third lowest amount in half a century.

And the uncertainty is growing just as the treaty approaches an important milestone: 2014 will be the first opportunity for Canada or the United States to announce that they plan to back out of the treaty or change it by the 60th anniversary of its ratification in 2024.

"It's significant," says Dr. Weber. "The Columbia River Treaty is a key international agreement for Canada."

▲ Work by the CFCAS-supported Western Canadian Cryospheric Network provided a first-ever inventory of Canada's almost 15,000 western glaciers and documented how fast they are changing—information critical to the water-flow forecasts used by BC Hydro, which generates 90 percent of its power from hydro dams.

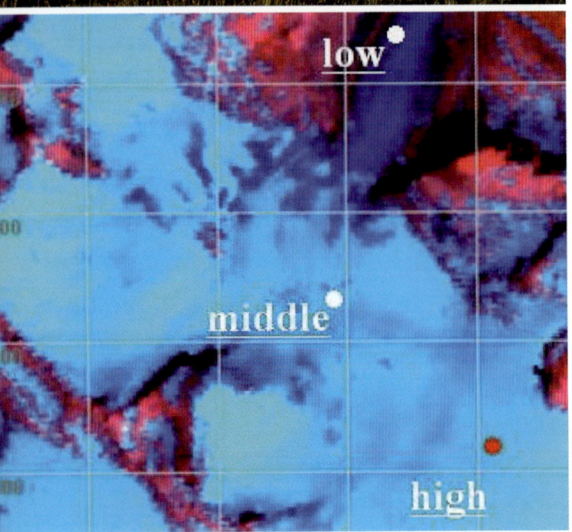

CANADA'S MOUNTAIN and northern glaciers are among the country's most spectacular and important land features. The future of these vast ice fields is critical to international obligations under the Columbia River Treaty, but they also play a role as a source of fresh water and renewable energy. They are vital to the survival of communities and forest ecosystems throughout the West and North.

In British Columbia, ice caps and ice fields cover about 100,000 square kilometres—almost one-tenth of the province's landmass—and glacier runoff helps generate the 90 percent of the province's electricity that comes from hydro dams. Glacial melt-water is also important for feeding and cooling the streams and rivers responsible for salmon, wildlife, and communities.

Many ice caps in the eastern Rocky Mountains, meanwhile, deliver badly needed late-summer irrigation water to Alberta farmland. And the more than 150,000 square kilometres of land glaciers in the Canadian Arctic are essential to the environment and to reflecting sunlight and mitigating warming. Yet, Canada's ice is melting.

In British Columbia, the area covered by glaciers shrank by 10 percent between 1985 and 2005, while the area covered by Alberta's ice sheets was down by a quarter in the same period. Overall, the rate of glacier ice-loss was about a half a percent each year across the region.

Meanwhile, water runoff from glaciers has been surging in Alberta, northwestern British Columbia, and the Yukon, adding measureable amounts to the world's rising seas. Farther south, the shrinking glaciers are now producing less water than they used to, limiting the water available for power dams.

"Really, it's a great unknown," says Brian Menounos of the University of Northern British Columbia. "Until recently, very few people were looking at this."

CFCAS-SUPPORTED RESEARCH has been tackling this "great unknown" for almost a decade. CFCAS networks and projects across the country have uncovered clues to the past, present, and future of glaciers in Canada—providing a first-ever glimpse at the extent and importance of the country's

ice sheets and the speed of changes caused by global warming.

The **Western Canadian Cryospheric Network** (WC2N), for example, is a five-year, $2.1-million scientific network headed by Dr. Menounos along with researchers from six Canadian and two American universities, as well as from government and private sectors. Using satellite and historic photographs as well as other clues, WC2N researchers have been able to piece together the previously unwritten story of the past, present, and future of the nation's western glaciers.

Before the work of WC2N, for example, no one had a clear idea of the size and coverage of Canada's estimated 15,000 western mountain glaciers. Network researchers were able to document the state of the glaciers today and, using evidence from the past, their size and extent going back almost 400 years.

The pace of melting they uncovered was alarming: glaciers in northwestern British Columbia and Alaska alone have lost about 42 cubic kilometres of ice every year for the past four decades, adding about five millimetres to ocean levels around the world and leaving the area covered by ice just three-quarters of the size it was 25 years ago.

The network researchers were able to use their understanding of past and present changes to model regional ice sheet changes into the future. These models are still being developed, but they already play an invaluable role for scientists and policy makers to anticipate how melting ice will affect the ability to generate power or deliver water from glacial run-off.

THE IMPORTANCE OF ICE—and the water it delivers—is not limited to the mountains of British Columbia. Work by other CFCAS-supported networks and scientific projects has revealed how changes to the great areas of land and sea ice elsewhere in Canada might also affect the water, weather, and climate vital to people and communities.

IP3 is the short name for a network called **Improved Processes and Parameterisation for Prediction in Cold Regions** (see Chapter 4). Work by IP3 researchers is documenting how land glaciers in the western Arctic and the northern

◄ View towards Mt. Kennedy from the Kluane icefields in Kluane National Park, Yukon (*far left*).

Melt water in a mountain glacial lake (*left*).

Revelstoke Dam, located on the upper Columbia River.

Rocky Mountains affect water cycles, weather, and climate in the region. The $2.5 million network involves scientists from eight universities, from governments, and from private industries across Canada.

IP3 research is being used by forecasters and policy makers to improve predictions about the availability of water from these northern and mountaintop sources. The forecasts are crucial to people living in the basins of the North Saskatchewan River, the South Saskatchewan River, and the Columbia River—the subject of the Canada-U.S. treaty and a source of water for many people living in western Canada and the United States.

Meanwhile, studies of land and sea ice in the Arctic are contributing to the work of the **Polar Climate Stability Network** (PCSN) as researchers uncover evidence of past changes to the climate and use this to model effects of receding northern ice on the climate of the future (see Chapter 4).

Other smaller CFCAS scientific projects have looked at glacial ice and water from other perspectives. CFCAS-supported researchers have used sediment layers from Upper Arrow Lake in British Columbia to study yearly runoff from the Columbia River over the past 1,000 years. Others have collected evidence of past climate changes trapped in the massive Ellesmere Island ice field, and another CFCAS-supported project revealed the extent of the climate-related shrinking of arctic lakes and ponds.

Shrinking glaciers

Glaciers and ice sheets store about **70 percent of the world's fresh water**, of which nearly 97 percent is stored in the large ice sheets of Antarctica and Greenland.

Mountain glaciers represent less than **0.5 percent of the total volume of ice** but supply fresh water to millions of people who heavily rely on their melt water.

The glaciers of western Canada and Alaska have retreated since the end of the Little Ice Age in the nineteenth century. The ice sheets stabilized and may have grown in the mid-20th century, **but resumed shrinking after 1980**.

Records of glacial change in the eastern Canadian Arctic are among the longest in existence, with many **covering more than 40 years**.

Between the 1960s and the mid-1980s, large arctic glaciers shrank only slightly. Most of the shrinkage affected smaller and more westerly glaciers.

The pace of glacial shrinking in the Arctic has **increased since the 1980s**.

Is hydro "green"?

The rushing water of streams and rivers—fed by rain or melting ice and snow—is often touted as Canada's answer to green energy. Few countries in the world share our potential for harnessing the flow of water to produce electricity—reducing our dependency on coal-fired or natural gas-fueled generating stations. In 2004, Canada was the world's largest producer of hydro power, and in British Columbia, Manitoba and Quebec, more than three-quarters of provincial electricity comes from hydro.

But just how clean and green *is* hydro? The question, it turns out, is a good one, and it was among those tackled by CFCAS-supported scientific projects exploring how hydro reservoirs can also become sources of methane—a potent greenhouse gas with 25 times the warming power of carbon dioxide.

A team led by Nigel Roulet of McGill University, for instance, examined the impact of Hydro-Quebec's recent Eastmain development and revealed the remarkable interaction between flooded land, soil carbon and other chemicals that generate greenhouse gas. In a separate CFCAS-supported project, Sherry Schiff at the University of Waterloo developed computer models of carbon cycles in reservoirs flooding different boreal forest types.

The work revealed that while reservoirs are responsible for greenhouse gas emissions, electricity produced from northern hydro reservoirs is nevertheless much less intensive than gas, oil or coal. The CFCAS-supported research was an invaluable first glimpse into the comparable impact of hydro on the climate.

6 THE LAND

B y any measure, 2009 was a difficult and costly wildfire year for British Columbia.

More than 2,440 forest fires burned through some 133,400 hectares of the province's woodlands. Dry weather and mature trees fed the furious flames. Thousands were forced from their homes, and at least three houses were consumed by the infernos.

The fires exasperated the Government of British Columbia, too. Costs for fighting the fires topped $209.5 million—almost three times more than the annual average.

"It has been a tough year," admitted British Columbia Forests Minister Pat Bell as the fire season drew to a close.

But as tough as it was on the provincial purse and British Columbia residents, it was a problem for the climate too—the fires released millions of tonnes of the greenhouse gas carbon dioxide, tipping the delicate balance of the forest carbon cycle and sending Canada's overall emissions upward.

"THE FACT IS, FORESTS ARE A BIG PART OF the carbon budget," says Hank Margolis of Laval University. "The net carbon exchange is a very small difference between two very large fluxes … and a small perturbation—fires or insects—can change the balance."

Carbon dioxide from Canada's more than 8,000 wildfires every year equals almost a fifth of all emissions from fossil fuels in this country. Insects also stir things up; the current mountain pine beetle outbreak in western forests is expected to send as much carbon skyward as five years of emissions from Canada's entire transportation sector.

▲ A group of friends watches wildfires burn on Terrace Mountain, north of Kelowna, B.C., in the early morning hours of Tuesday August 4, 2009. The evacuated community of Fintry is seen at right. Hundreds of firefighters battled raging forest fires in British Columbia.

◀ Muncho Lake, in British Columbia. The deep blue colour of this lake, known to locals as "blue lake," is due to the absence of glacial feed. This is pristine Rocky Mountain spring water, not melted glacial ice.

▼ Mountain pine beetle.

These natural disturbances get their significance by tipping a balance in Canada's vast tracts of forest. Seven percent of the planet's woodlands are here. The soils, plants, wood, and bogs of these forests are an enormous storehouse of carbon, cycling millions of tonnes of carbon dioxide greenhouse gas back and forth between the atmosphere and the land.

Importantly, this cycle can shift quickly from absorbing more carbon from the air through photosynthesis (acting as a carbon "sink") or releasing it (as a carbon "source") through decay, respiration, or burning—a change that can have a measureable impact on greenhouse gases in the atmosphere and changes to climate around the globe.

"It's important from a stewardship perspective to understand what's going on with our carbon because we have so much of it and it has such a potentially big impact on the rest of the world," says Dr. Margolis.

FOR 10 YEARS, CFCAS-supported research networks and projects have helped to understand how Canada's vast, verdant landscape contributes to or absorbs the greenhouse gases that are altering our climate. The work has documented the exchange of gases and energy in different regions across Canada and studied the effects on the atmosphere of biological processes and disturbances, such as fires, insects, or forest harvesting.

Dr. Margolis, for example, is the lead investigator with the four-year, $4.5-million **Canadian Carbon Program** (CCP), a CFCAS-funded research network that has detailed how Canada's forests and peatlands "breathe" airborne carbon

Scientists say most of Canada's forest acted as a net sink for greenhouse gases for two years in 2000 and 2001 before becoming a net source in 2002. It is expected to continue to release more climate-warming gas than it absorbs through 2022.

dioxide. The CCP network involves dozens of scientists and graduate students from 15 universities across the country, as well as researchers from federal and provincial government agencies.

The CCP picks up and expands on earlier research by the five-year, CFCAS-supported Fluxnet Canada network, which began the work of building and equipping towers to measure carbon exchange over forests and bogs across Canada.

The "flux" towers use sophisticated gas analyzers and wind-speed meters mounted high over the forest canopy to calculate the exchange of carbon between natural ecosystems and the surrounding air. Set up at as many as 12 sites in different forests and peatlands across the country, the towers provide almost continuous, year-after-year information about how carbon is absorbed or lost back to the atmosphere and how warmer temperatures and major disturbances—fires and insects, for example—can alter this important cycle.

"The idea behind Fluxnet Canada and the Canadian Carbon Program was to get an idea of the role of Canadian forests in the global carbon cycle," says Dr. Margolis. "To what extent are they sinks or sources? What makes them sinks or sources?"

The filtering forest

Canada's vast northern forests cover almost half of the country and play a key role in regulating our climate, but Frank Wania at the University of Toronto knows that woodlands don't have to be immense or remote to be important to our environmental health.

For three years beginning in 2001, Dr. Wania and his team used CFCAS support to study how woods and trees operate as biological filters for air pollution produced by nearby cities and industrial developments. The scientists documented how mainly deciduous forests absorbed certain toxic air pollutants—including dangerous PCBs, or polychlorinated biphenyls—from the atmosphere and then transferred them to the forest floor with the falling leaves.

Dr. Wania, along with Tom Harner of the Meteorological Service of Canada, revealed just how important forests and woods can be to significantly reducing the air pollution that puts people and wildlife at risk.

The tower data is helping researchers understand what's going on within forests and fens to influence carbon, energy, and water changes. It is also providing the tools scientists need to model carbon exchange on a larger scale and to show how this is influenced by atmospheric shifts, temperature, and other factors.

OTHER CFCAS-SUPPORTED research has also studied the cycling of carbon and its influence on climate. For instance, an earlier CFCAS-supported network used computer simulations of carbon exchange between the land and the air to help predict change into the future.

The **Canadian Global Coupled Carbon Climate Model**—known by the abbreviation CGC3M—was a five-year, almost-$2 million network that tested and refined computer-built mathematical models of interactions between carbon dioxide, forests and other living landscapes, and the climate. The model has since become a key part of larger, so-called "coupled" models of climate that help researchers better predict atmospheric and climate changes down the road.

Meanwhile, several smaller CFCAS-supported projects have also been exploring important interactions between the Canadian landscape and our atmosphere and climate.

For example, researchers from across the country have used CFCAS funds to explore and forecast what will happen as the changing climate warms the country's northern permafrost—areas of permanently frozen ground. The work has documented for the first time what these changes might mean to the present and future carbon cycling role of forest soils, farm fields and mountain vegetation.

CFCAS projects have also given us a better understanding of past shifts in Canada's terrestrial carbon cycle—as well as the history of the land-atmosphere exchange of other greenhouse gases, such as methane and nitrous oxide. Studies of soil organic carbon, meanwhile, have improved our understanding of how future warming will affect the large stores of carbon that exist underground.

Canada's green carbon storehouses

Canada's managed forest makes up **13 percent of the land area in North America** and, in the 1990s, contributed only two percent of the carbon "sink."

The importance of forests in the global carbon cycle is one reason why the **Kyoto Protocol** gave countries the option of including forest management in their Kyoto accounting for 2008-2012.

Peatlands store more carbon by equivalent area than any other terrestrial ecosystem and cover about 17 percent of Canada's land area.

The carbon-storing capacity of global forests could be lost entirely if the **earth heats up 2.5° Celsius above pre-industrial levels**, according to the International Union of Forest Research Organizations (IUFRO).

Canada's boreal forests (as well as those of Finland, Russia and Sweden) are expected to warm more than forests in the tropics, and although warmer temperatures could expand forests northward, the short-term positive impacts would be cancelled out by increased insect invasions, fires, and storms.

The CFCAS-supported Canadian Carbon Program is supplying detailed information about the carbon dioxide ▶ exchange between Canada's forests and the atmosphere and providing important scientific insight into how forest carbon offset policies could be used in possible international "cap-and-trade" carbon reduction plans.

7 THE OCEANS

W

hen Hurricane Juan rolled off the Atlantic and slammed into Halifax on September 29, 2003, no one was expecting the most devastating windstorm in more than a generation.

"This storm has kind of surprised us," said a startled Mike Myette of Nova Scotia's Emergency Measures Organization, as winds of more than 150 kilometres an hour tore into the city. "A lot of time we see hurricanes peter out before they hit land."

Meteorologists had been tracking Juan for days and had downgraded it to the weakest hurricane category—Category 1—just a day before it came ashore. Residents had been warned, but everyone expected little more than just another strong Atlantic blast. What arrived instead were fearsome Category 2 winds and towering 18-metre waves. Strangely, the hurricane also brought the destructive impact of a far stronger, Category 3 storm.

Juan smashed windows and tore at roofs. It tossed fishing boats ashore and buckled piers. The high winds, waves, and rain caused an estimated $200 million in damage, and millions of trees across the province were toppled. Eight people died, and hundreds were evacuated. About 300,000 residents were without power across the province—many for almost two weeks.

Since the storm, many Nova Scotians have wondered how forecasters and others got it so wrong. The hurricane's unexpected ferocity was another example of how complicated—and unpredictable—ocean-born storms can be. Now, as the changing climate transforms the North Atlantic into an even more potent storm factory, the question of the faltering forecast that preceded Hurricane Juan may be more relevant than ever.

▲ A resident of Bedford, Nova Scotia, walks past a boat that washed ashore in the front yard of a home in the aftermath of Hurricane Juan. Two people were dead and 80,000 were without power in Nova Scotia the morning after Hurricane Juan barreled into the eastern Canadian province from the Atlantic Ocean. Juan hit Halifax, Nova Scotia's biggest city and one of the country's largest ports, with winds of up to 140 kph (90mph). A state of emergency was declared in the city.

"MEMORIES OF JUAN [ARE] STILL FRESH," says Hal Ritchie, an Environment Canada scientist and adjunct professor at Dalhousie University.

Juan was one of 14 named storms and among three Atlantic hurricanes to hit Canadian shores after brewing over the ocean between 2001 and 2008. Several other storms—such as Hortense (1996), Michael (2000), Gustav (2002) and Kyle (2008)—also landed with hurricane force along Canada's east coast in recent years.

Scientists believe that a warmer sea surface and other ocean effects from climate change are affecting coastal storms. Some hurricane researchers suggest that damage from a growing number of very fierce—Category 4 and 5—hurricanes on the Atlantic is likely to outweigh the benefits of fewer less-intense storms in a warmer future. But the real impact of global warming on marine storms is still not clear.

Researchers have known for a long time that oceans and their powerful globe-spanning currents are intimately linked to our climate and weather, including hurricanes. But for years, understanding the detailed nature of that relationship remained elusive. The knowledge gap often frustrated forecasters, and it meant models of the atmosphere that researchers used to understand and predict climate change did not include many of the important influences of oceans.

CFCAS-SUPPORTED RESEARCHERS have spent the past decade revealing just how close the links are between oceans and the complex machinery of our climate and atmosphere. In particular, CFCAS research networks and projects have used tools as varied as satellite images and data from thousands of instrument-equipped buoys to describe ocean systems in unprecedented detail, and to model their interaction with changes overhead.

The CFCAS-supported **Global Ocean-Atmosphere Prediction and Predictability** (GOAPP) **network**, for example, is a $3-million scientific network headed by Dr. Ritchie and his Dalhousie University colleague Keith Thompson.

For four years, the network has linked ocean-climate interactions through its unique cross-country collaboration among more than 40 scientists, researchers, and students representing two distinct and typically isolated disciplines—oceanography and atmospheric science.

The result has been an opportunity to transform short- and long-term marine-weather forecasts by tying in—or "coupling"—weather models with a more accurate picture of changes taking place on the sea surface.

These changes are measured across areas too wide and time-scales too large for scientists to anticipate the arrival of hurricanes. However, the research is expected to improve the prediction of large, fierce, and lasting storms that mean danger

An iron fix is scrapped

Can iron-fortified oceans solve climate change? The question has long generated hope for climate watchers and others looking for a quick fix to global warming—or, at least, an easy way to earn carbon credits in greenhouse gas offset programs. Iron stimulates growth in photosynthetic plankton. And these tiny organisms—drifting in countless numbers in the sea—absorb the carbon dioxide that contributes to global warming.

Thanks to the work of scientists with the CFCAS-funded **Canadian Surface Ocean-Lower Atmosphere Study** (C-SOLAS) network the idea can finally be laid to rest—or rust.

C-SOLAS was a $4.4-million, five-year network headed by Maurice Levasseur of Laval University and before him, by Bill Miller, then at Dalhousie University. It represented Canada's contribution to an international effort (known as SOLAS) to better understand how changes in the living and non-living elements in oceans—known as the ocean biogeochemistry—respond to climate change and, in some cases, affect it.

The network studied air-sea biogeochemical interactions in the northwestern Atlantic Ocean; it also conducted an unprecedented large-scale experiment to find out what would happen if iron was dumped into the North Pacific. The research benefited from strong collaboration between Fisheries and Oceans Canada, Environment Canada and the academic community. It was groundbreaking and showed for the first time that iron fertilization of oceans is both inefficient and potentially harmful to the long-term stability of the Earth's climate system. The iron also risked changing natural ocean ecosystems and stimulating the production of other greenhouse gases such as nitrous oxide.

Using both ocean and atmosphere data together to generate real-time forecasts for Canada's East Coast made the CFCAS-supported Lunenburg Bay network a world leader in marine weather prediction. Network findings and information were readily available and accessible to anyone using the Internet; they could even be downloaded on a cell phone.

for fishers, sailors, and other Canadians living or working along the nation's sea coasts. In fact, GOAPP research has been a key piece in a federal government forecasting initiative called the Canadian Operational Network of Coupled Environmental Prediction Systems (CONCEPTS).

By "coupling" a growing knowledge of ocean systems to existing large-scale climate models, GOAPP researchers are also helping scientists, including those involved in the Intergovernmental Panel on Climate Change, to track and predict global climate change.

"The ocean provides the memory that holds the hope for greater predictability by connecting the ocean and atmosphere together," says Dr. Ritchie, who also works with Environment Canada's Atlantic Environmental Prediction Research Initiative.

GOAPP research spans the reach of the world's ocean basins and currents through the network's participation in a 23-nation ocean-monitoring research program known as **Argo**. The sophisticated instruments aboard Argo's 3,000 research buoys floating throughout the planet's seas reveal surface and underwater conditions and transmit the data to land every 10 days, via satellite.

GOAPP IS NOT THE ONLY CFCAS-funded research network that is simultaneously studying ocean-climate interactions and playing a key international role in global ocean science.

For example, an earlier CFCAS-supported network known by the acronym C-SOLAS provided Canada's contribution to the global **Surface Ocean–Lower Atmosphere Study** (see Sidebar: An iron fix is scrapped). The network, based at Dalhousie University and involving dozens of academic and federal researchers, and graduate students from universities across the country, was a world leader in understanding how living and non-living elements in oceans affect and respond to changes in the global climate.

The network demonstrated the extraordinary achievements attainable through the shared efforts of funding agencies, federal departments—in this case Fisheries and Oceans Canada,

and Environment Canada—and the academic community. C-SOLAS was the first SOLAS national program to receive funding and as such, played a leading role in the growth of international SOLAS efforts in more than 20 countries.

Similarly, several research projects of the CFCAS-supported **Polar Climate Stability Network** (PCSN; see Chapter 4) have studied the past and present role of sea-sky interactions in the Arctic Ocean on climate variation. The work provided detailed information used to improve models and to better predict the response of the Arctic and elsewhere to climate warming into the future.

Meanwhile, a previous CFCAS-supported research network at Lunenburg Bay—called the **Interdisciplinary Marine Environmental Prediction in the Atlantic Coastal Region** network, or simply the **Lunenburg Bay network**—focused its attention on creating a real-time system for using ocean and atmosphere information to predict the often temperamental and sometimes dangerous weather of Canada's Maritimes.

Using an array of automated environmental sensors in Lunenburg Bay, the five-year, $4.7-million network collected continuous observations that helped researchers improve computer models capable of predicting sea and weather changes that might generate coastal flooding or ocean surges.

Many smaller CFCAS-funded projects have also improved the science behind marine forecasting, making the behaviour of ocean-generated storms and other weather easier to predict. For example, researchers used support from CFCAS to help describe everything from the influence of whitecaps and breaking waves on air-sea interactions to how Atlantic hurricanes transform themselves during their northward journey from the tropics. CFCAS projects also explored the impact of equatorial El Niños on Canadian winter weather forecasts thousands of kilometres away, or the effects of underwater waves and ocean mixing on marine weather prediction.

Other projects have focused more on the role of oceans as an influence on climate change or vice versa. For instance, CFCAS-supported scientists have studied how the increasing acidity of oceans—caused by more carbon dioxide—affects sea life or how melting arctic ice influences the ocean circulation important to climate. Some of the research has described how the deep water under Antarctica works as a reservoir of greenhouse gas or how iron in the Arctic Ocean affects the carbon cycle in the Far North. Other work documented 2,000 years of past climate-ocean transformations off the coast of British Columbia to improve our understanding of future change.

The oceans and the climate

The oceans are getting warmer. Between 1961 and 2003, the average global ocean temperature has risen by 0.10° Celsius down to a depth of 700 metres.

Rising levels of carbon dioxide in the air are changing the chemistry of the seas. The inorganic carbon in world oceans has increased since the beginning of the industrial revolution (about 1750) and it continues to rise.

In recent decades, **more than a third of all carbon dioxide emissions** into the atmosphere have been absorbed by oceans—down from an average of more than 40 percent of emissions during the period between 1750 to 1994.

The airborne carbon dioxide absorbed by oceans has **made the seas more acidic** by an average ocean pH of about 0.1 since 1750; the acidity is bleaching and killing the coral that sustains many fish ecosystems, and it is preventing shells from hardening for many shell-forming organisms.

Between 1993 and 2005, **sea levels around the world rose by almost 3.3 millimetres every year**, according to recent estimates using satellite measurements. Some researchers suggest oceans may rise as much as a metre by 2100.

The number of major **Atlantic hurricanes** per year is expected to almost double by the end of the century in response to global warming.

Damage from a larger number of **very strong — Category 4 and 5 — hurricanes** is likely to outweigh an expected decline in less-intense storms.

8 THE SKY

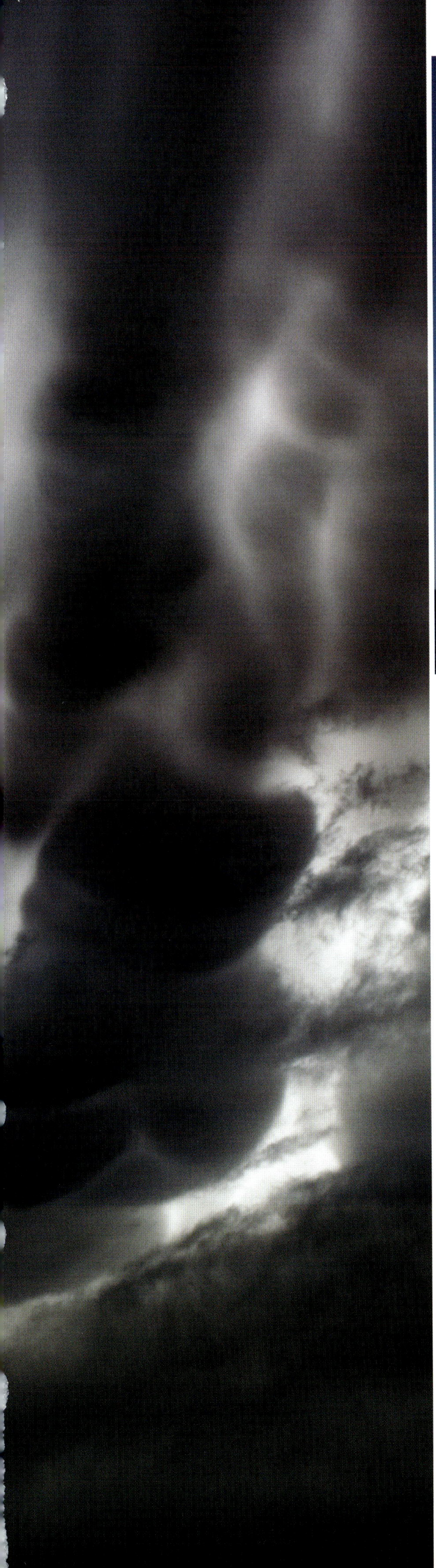

A shortage of sunshine in Canada is costing lives and billions of health care dollars, according to a recent study by doctors and nutritionists. And now, thanks to climate change, it's likely to get worse.

Exposure to sunlight and its ultraviolet (UV) rays helps skin generate vitamin D—the so-called "sunshine vitamin" that's been hailed for preventing a host of afflictions from bone disease to cancer. But Canadians don't get enough of it during dark, cold winters when we often huddle indoors. According to Statistics Canada, two-thirds of Canadians have less-than-ideal vitamin D levels, while one in 10—more than three million people—have amounts low enough to be considered unhealthy by doctors.

A 2010 study suggests up to 37,000 lives are lost and an estimated $14.4-billion is spent on diseases related to the low vitamin D levels of Canadians. "The results of this study strongly suggest the personal and economic burden of disease in Canada could be significantly reduced if optimal vitamin D levels are increased," said Dr. William Grant, one of the study's authors.

Unfortunately, our changing climate and its effect on sun-filtering ozone in the atmosphere will not make that prescription any easier to fill: global warming is causing more ozone to be distributed over Canada.

"Ozone helps reduce UV radiation on the ground. So, by changing the distribution of ozone, climate change could actually make UV go up or down depending on the region," says Ted Shepherd, an atmospheric scientist at the University of Toronto. "Over Canada, it looks like it will actually go down and not up. That might be good for sunburns, but it turns out that it's actually really bad for vitamin D production in the body."

▼ A research collaboration supported by CFCAS has identified which airborne particles are most damaging to human lungs, providing valuable information to governments and agencies that regulate air pollution and helping to reduce billions of dollars of smog-related health care costs every year.

THE RELATIONSHIP BETWEEN ultraviolet radiation, ozone, and climate demonstrates how complicated our sky can be. What happens many kilometres above our planet's surface is not only complex and difficult to predict, but it also has impacts that can reach around the world—or sometimes down to Earth where people's lives, livelihoods, and even their health are affected.

Many scientists studying climate focus their attention on the curious systems and gassy processes of the lower regions of the atmosphere (the troposphere)—where our daily weather happens and much of the air's water vapour and gas particles are found. But the sky above the troposphere is also an important climate player. And the chemistry and turbulence that take place in the so-called stratosphere—between 10 and 50 kilometres up—are also influential.

Most of the atmosphere's ozone gas, for instance, is found in the stratosphere (commonly called the ozone layer). Ozone protects us (and everything else) from 90 percent of the sun's searing and cancer-causing ultraviolet rays. In the 1970s and 1980s, dwindling atmospheric ozone was linked to a rise in airborne concentrations of a group of industrial chemicals called chlorofluorocarbons (CFCs) used in refrigerators and spray cans. A worldwide ban on CFCs—known as the Montreal Protocol—was signed in 1987.

FOR THE PAST DECADE, CFCAS-supported researchers have been searching the skies for answers to some of the most vexing high-altitude questions about our climate, our weather, and even the quality of our air. The work has revealed valuable insights not only into future global change but also into the behaviour of airborne chemicals and pollutants that directly affect the health of Canadians.

For example, the CFCAS-funded network known by the acronym **C-SPARC** has spent the past four years studying changes to ozone and other chemicals adrift in the stratosphere.

The network, headed by Dr. Shepherd and involving more than two dozen researchers from universities and government agencies across the country, is a leader for the larger international World Climate Research Programme initiative exploring **Stratospheric Processes and their Role in Climate** (SPARC). Indeed, the SPARC International Project Office has been based at the University of Toronto since 2004.

Groundbreaking work by C-SPARC—a four-year, $3.4-million scientific network—was among the first research to detail clearly how changes in the chemistry and circulation high up in the stratosphere affect climate change and ozone. The findings have been vital for improving computer models of climate and ozone that previously omitted this vital information.

The research—along with work by a similar, earlier CFCAS-supported network called **Modelling of Global Chemistry**

El Ninõ climate events have their origins in the tropical Pacific, but they are nevertheless capable of causing millions of dollars of weather havoc to Canadian farms, forests, and fisheries. A CFCAS-supported project at the University of Northern British Columbia has created new tools to better predict El Ninõs and to anticipate their impact on far-away Canada.

Blowin' in the wind

Air—as Allan Bertram of the University of British Columbia knows—is a restless global citizen; it goes anywhere and the pollution it carries has little respect for national borders.

Air is also the stuff we breathe, and its quality can have serious consequences for health and for the environment, including climate. Most Canadians count on federal or provincial governments to regulate polluters and to stop any dangerous emissions that might make us sick. But regulations have little impact on the particles and dust carried on prevailing winds across the northern Pacific.

Dr. Bertram and his colleagues have focused their two-year CFCAS-funded project on understanding the chemistry of air that reaches western Canada from far-off Asia—a place where many countries are ramping up industrial development and economic growth and relying mainly on coal and other fossil fuels to do it.

The researchers modified a tool known as a mass spectrometer to build a specialised instrument that can quickly discern the make-up of tiny airborne particles. The instrument was installed high on the top of British Columbia's Whistler Mountain—the renowned site of the 2010 Olympic downhill ski races and other events. There, at a weather station operated by Environment Canada, the spectrometer takes year-round measurements of the chemical composition of drifting air without the confusing signals of local air and pollution whirling around.

Dr. Bertram, along with other researchers from the University of British Columbia, Environment Canada, and the Pacific Northwest National Laboratory in the United States, have used the data to identify air particle signatures that can pinpoint the origin of globe-trotting pollutants. They have also described how chemicals, such as airborne sulphur, change local climate by preventing dust and water vapour from coming together to form rain clouds.

The work has provided important air quality information for forecasters and for policy makers who want to lessen the effects of air pollution in Canada—no matter how well-travelled it might be.

for Climate—has refined our ability to understand and predict future global change, but it has also shed light on what the post-Montreal Protocol recovery of the ozone layer might mean in a warming world.

Dr. Shepherd's work, for instance, was the first to reveal how the climate-caused redistribution of ozone will likely spell more bad news for vitamin-D deficient Canadians.

"High latitude countries like Canada already have a problem because they don't get enough sunshine," says Dr. Shepherd. "So, there are all sorts of health implications."

MANY OTHER CFCAS-SUPPORTED networks also have their eyes on the sky. Climate modelling efforts, for instance, have been an important focus for CFCAS-funded research. Chief among these was the five-year, $2-million network for the **Development of a Canadian Global Coupled Carbon Climate Model** (CGC3M) which developed and tested an integrated model that incorporated the relationship between the global carbon cycle and climate (see Chapter 6).

Meanwhile, another five-year, $2.3-million initiative known as the **Climate Variability: Causes and Predictability** (CLIVAR) network developed new analytical tools for understanding how much global warming is due to natural variability compared to human causes. Two other CFCAS-supported efforts—the **Canadian Regional Climate Modelling and Diagnostics** (CRCMD) network and the earlier **Canadian Regional**

Climate Modelling network—developed models that characterized the distinctive regional features of climate change.

To better understand the climate influence of clouds and airborne particles, two CFCAS-supported networks—the **Cloud-Aerosol Feedbacks and Climate** (CAFC) network and the **Modelling of Clouds and Climate** (MOC2) network—linked with international efforts and carefully modelled the relationship between clouds and climate.

Air quality is another area of concern for CFCAS-supported scientists exploring the intricacies of the sky. The **Canadian Network for the Detection of Atmospheric Change** (CANDAC; see Chapter 4), for instance, used data collected just 1,100 kilometres from the North Pole to reveal the atmospheric effects of air quality, climate, and ozone.

The CFCAS-supported **Multiscale Air Quality Modelling network** (see Chapter 3) developed a state-of-the-art online model of air quality for use by the Meteorological Service of Canada and provincial ministries of the environment. Meanwhile, the work of the **Pacific 2001** network (see Chapter 3) in British Columbia's Fraser Valley helped federal scientists pave the way for new policy guidelines on air quality.

Smaller CFCAS-funded projects have supported climate and atmospheric scientists with interests as varied as soot, past temperature changes, ice clouds, ozone flux, water vapour in the tropics, arctic aerosols, El Niño effects, the predictability of atmospheric change, and other features of coupled climate modelling.

9 THE WORLD

O

ur changing climate "is, simply, the greatest collective challenge we face as a human family," said United Nations Secretary-General Ban Ki-moon during a 2009 keynote speech in Seoul, South Korea. Countries of the world need to work together not only to slow change and adapt to it, but—first and foremost—to understand it.

For decades, scientists around the planet have joined forces to piece together some of the most intimidating global climate puzzles. The success of their work—and its importance to climate policy decision makers—owes much to the coordinating efforts of world climate organizations and the willingness of nations to participate in them.

Since the middle of the last century, for instance, climate scientists and the governments that support them have rallied around international groups such as the World Meteorological Organization (WMO), the World Climate Research Programme (WCRP), and the Intergovernmental Panel on Climate Change (IPCC). From these groups, ground-breaking collaborative scientific projects have come together, such as the Stratospheric Processes and their Role in Climate project (see Chapter 8) or the Surface Ocean-Lower Atmosphere Study (see Chapter 7).

For more than 50 years, these international efforts have shared a common and notable feature— the significant presence and frequent leadership of Canadians.

"Canada has always punched way above its weight in the international scientific community," says Tim Aston, science officer at CFCAS headquarters in Ottawa. "I think we can safely say it's the same with respect to climate science."

▲ U.N. Secretary-General Ban Ki-moon talks during a news conference after the 27th Plenary of the Intergovernmental Panel on Climate Change (IPCC) in Valencia, Spain, November 17, 2007. Sitting next to him is IPCC Chairman Rajendra Pachauri (*second left*), Secretary-General of the World Meteorological Organization Michel Jarraud (*left*) and United Nations Environment Programme Executive Director Achim Steiner (*right*).

▲ The Polar Environment Atmospheric Research Laboratory (PEARL), Nunavut: an ideal spot to study Earth's protective ozone layer.

IN A COUNTRY WHERE mercurial weather and climate make a profound difference to the daily lives of Canadians, climate and atmospheric research has always been a natural fit. Its relevance in Canada provides a sense of local urgency and significance.

But changes in Canada—with its vast tracts of carbon-cycling forests and fields and its climate-critical arctic sea ice—also have a profound influence on changes happening elsewhere. Canadian scientists are perhaps uniquely equipped to understand both the global nature of climate questions and the international need to find answers.

"I think we realize that climate issues do not stop at geographical boundaries," says Dr. Aston. "I think this is a paradigm shift we are starting to see in some other countries now as well: the recognition that national-level resources need to be directed at international-level concerns."

CFCAS AND CFCAS-FUNDED climate scientists have always been at the vanguard of this change in thinking. From the start, CFCAS recognized the interconnected complexity of world climate, oceans, and weather; efforts to understand global change meant keeping the big, international picture in focus.

A world of science support

The global reach of climate, weather, and oceans means more scientists from more countries are recognizing the need to work together to find answers. The organizations from around the world that support these researchers are also coming to the same conclusion.

International recognition resulted in CFCAS leadership of the International Group of Funding Agencies for Global Change Research (IGFA) from 2007-2010. This consortium of funding agencies from more than 20 countries has played a key role in encouraging and supporting work on international climate science priorities and in identifying opportunities to coordinate the global climate science efforts that make a difference to governments and to policy makers around the world.

CFCAS has served as IGFA's global secretariat and CFCAS Executive Director Dawn Conway as its Chair. Work by the Canadian-led partnership has been crucial for exchanging information on national research programs, global initiatives, emerging issues and "best practices." CFCAS has also helped IGFA members explore international funding opportunities.

In 2008, the CFCAS executive director coordinated IGFA feedback on two international program evaluations and served on the international review committee for the Earth System Science Partnership, based in Paris. The review committee's report was released in mid-2008. Ms. Conway and CFCAS Science Officer Tim Aston have chaired important IGFA committees, and CFCAS hosted the 2006 IGFA meeting in Montreal.

That's why, for example, the CFCAS Board of Trustees agreed to serve as the Canadian National Committee (CNC) for the World Climate Research Programme. The WCRP is a key collaborative initiative of the International Council for Science, the World Meteorological Organization, and the Intergovernmental Oceanographic Commission of UNESCO. It is a vital worldwide organization charged with improving climate prediction and with better understanding our human influence on climate.

Through the encouragement and support of CFCAS and the CNC, Canadian climate scientists have been instrumental in many WCRP projects, including research projects exploring everything from water availability and cold climates to atmospheric processes, climate shifts, and the ocean surface. The National Research Council of Canada (NRC) has relied on CFCAS to report on Canada's leadership and participation in the WCRP and on the organization's benefits to Canadians.

CFCAS has also been an important supporter of Canada's national secretariat for International Polar Year (March 2007 to March 2009). IPY is a 60-nation international scientific program organized through the International Council for Science and the World Meteorological Organization. IPY stimulates and highlights research achievements in the Arctic and the Antarctic. Canada will host the final IPY international conference in Montreal in April 2012.

On a smaller scale, CFCAS funding also helped leverage support to create other world-class scientific institutions. In 2002, for example, Hugo Beltrami of Nova Scotia's St. Francis Xavier University used a CFCAS grant to stimulate the creation of the Environmental Sciences Research Centre (ESRC). The centre encourages partnerships with scientists—and, importantly, graduate and postdoctoral students—from across Canada, as well as from Spain, Romania, Germany and the United States, to study clues to climate and permafrost history that are trapped in the ground.

CFCAS RECOGNITION OF the international scope of climate science has also meant that many CFCAS networks and projects have become key components of larger global initiatives. For example, CFCAS-supported contributions to the worldwide **Surface Ocean–Lower Atmosphere Study** (SOLAS; see Chapter 7) helped to reveal new and important science around how the air and oceans interact and how the chemistry of the sea (concentrations of iron, for example) could affect the future of our climate.

Meanwhile, the International Secretariat for the **Stratospheric Processes and their Role in Climate** project (SPARC; see Chapter 8) was a clear example of Canada's remarkable climate science profile in the world. The enormous multinational effort to understand how changes in the stratosphere affect global climate was based at the

CFCAS support put Canada in the lead of international efforts to understand how the middle atmosphere affects climate and ozone. A CFCAS-funded coordinating office served as the International Secretariat for the **Stratospheric Processes and their Role in Climate** (SPARC) program, with headquarters at the University of Toronto.

University of Toronto and supported by CFCAS, Environment Canada, the Canadian Space Agency, and the university itself.

Many other CFCAS networks, such as the **Drought Research Initiative** (see Chapter 2), the **Polar Climate Stability Network** (see Chapter 3), and the **Western Canadian Cryospheric Network** (see Chapter 5), included university collaborators from other countries, reflecting the global implications of much of the Canadian-led research.

NOT ONLY ARE CFCAS-SUPPORTED scientists "punching above their weight," many are being celebrated for it. For instance, several CFCAS-associated researchers are among those who, as part the international scientific team of the Intergovernmental Panel on Climate Change, shared in the honour of the 2007 Nobel Peace Prize awarded jointly to the panel and former U.S. vice-president Al Gore.

Several lead investigators with CFCAS-supported networks have other national and international honours, and many also hold prestigious Canada Research Chair positions at their universities.

Canada on the world's climate science stage

1951 | Canada is part of the World Meteorological Organization (WMO), a specialized agency of the United Nations.

1979 | Canada participates in WMO's First World Climate Conference.

1980 | The WMO and the International Council for Science (ICSU) sponsor the creation of the World Climate Research Programme (WCRP).

1987 | Canada and other world nations sign on to the Montreal Protocol on Substances that Deplete the Ozone Layer.

1988 | The Intergovernmental Panel on Climate Change (IPCC) is established.

1997 | Canada signs on to the multinational Kyoto Protocol of targets and timetables for reducing greenhouse-gas emissions.

2004 | The International Secretariat for a WCRP worldwide research project studying Stratospheric Processes and their Role in Climate (SPARC) moves to Toronto.

2007 | Canadian scientists are key participants in the IPCC and share the honour of its Nobel Peace Prize, awarded jointly to the IPCC and former U.S. vice-president Al Gore.

QUESTIONS REMAINING

"THERE IS SO MUCH TO LEARN, to be discovered, and to be transformed into effective actions for Canadians," says Gordon McBean, chair of the CFCAS Board of Trustees and a celebrated climate scientist.

"Without sound scientific information, how will the government evaluate the effectiveness of green technologies, or build northern infrastructure, or develop our energy industry, or assure water supply and clean air?"

For 10 years, CFCAS has been helping the best and brightest Canadian scientists work together to find answers to questions like these—answers that governments, industries, and other Canadians need to grasp the opportunities and to meet the challenges of a fast-changing world.

IN JUST ONE REMARKABLE DECADE, CFCAS has launched and supported 198 scientific networks and projects that have successfully tackled important questions concerning our sky, our environment, and our future. But, perhaps most importantly, these initiatives have solved weather and climate mysteries that matter on the ground—right now, to all Canadians, and in regions all across the country.

Work supported by CFCAS—Canada's premier funding agency for all university-based climate science—has helped millions of Canadians anticipate city heat waves and protect their health; it has helped prairie farmers prepare for lasting, billion-dollar droughts; it has provided tools for regulating the air pollution that's affecting our lungs; and it has improved marine weather forecasts for coastal communities often battered by ocean storms.

"To plan for the future, people need to know how their local conditions will change, not how the average global temperature will climb," argued a 2010 commentary in the leading scientific journal *Nature*. In other words, the important work of climate science researchers is still being done "to develop tools to accurately forecast climate changes for the twenty-first century at the local and regional level."

Ten years of work by CFCAS has helped make Canada a world leader in understanding climate and weather. But the Foundation's focus has also been on the regional and even local needs of businesses, farmers, communities, and Canadians throughout the country's many and varied environments. It is in these places where the uncertainties of our climate future matter most—and it is also where many questions remain.

THE REMARKABLE ACHIEVEMENTS OF 10 YEARS OF CFCAS-supported science are worth celebrating. Thanks to CFCAS, Canadian researchers have played an internationally significant role in understanding how the world's weather and climate are changing and why, and they have influenced the direction of international efforts. But the work of CFCAS-funded scientists is far from done when it comes to understanding the atmospheric, oceanographic and hydrological conditions at regional and local levels.

Canada is expected to experience more change than many other countries as the world grows warmer. These shifts will have dramatically different effects whether you live in the Far North, in central Saskatchewan, or along Halifax's famous waterfront.

Change can bring opportunity, but only when we possess the knowledge and skill to recognise it and to take advantage of it. For 10 years, CFCAS has helped Canada seize opportunities for future prosperity as well as adapt to and slow climate change—but the job is far from finished.

Research is the only antidote to the continuously-growing number of weather and climate questions that affect Canadians today and in the future. Only the central science support and direction of an agency like CFCAS can ensure that university scientists working in meteorology, oceanography, and atmospheric sciences have the stable funding environment they need to keep up the critical hunt for answers.

"It's really important that Canada's efforts are organized into research
networks as CFCAS has done so we have compatible data and
a coherent network that can interact with the rest of the world."

LAVAL UNIVERSITY SCIENTIST HANK MARGOLIS,
LEAD INVESTIGATOR WITH THE CANADIAN CARBON PROGRAM (CPP)

"There has never been an opportunity like this to look at drought
in a comprehensive manner before. Without CFCAS, this simply wouldn't
have happened."

UNIVERSITY OF MANITOBA SCIENTIST RON STEWART,
LEAD INVESTIGATOR WITH THE DROUGHT RESEARCH INITIATIVE (DRI)

"This project wouldn't have come about without the help of CFCAS. We
might have seen some of the model development, but not the evaluation to
go with it…. But there's a bigger picture. The CFCAS money is important
because it's training a generation of people that will look at urban-scale
environmental issues, allowing us to keep this area of research at the fore in
all the universities involved and in some ways within Environment Canada.
They see a lot of the benefits, too."

UNIVERSITY OF WESTERN ONTARIO SCIENTIST JAMES VOOGT,
LEAD INVESTIGATOR WITH THE ENVIRONMENTAL PREDICTION
IN CANADIAN CITIES NETWORK (EPICC)

"The PCSN's strength is in the additional power that we developed through
the collaborative process of the network of scientists. …It wouldn't have
been possible without the CFCAS."

UNIVERSITY OF TORONTO PHYSICIST RICHARD PELTIER,
LEAD INVESTIGATOR WITH THE POLAR CLIMATE STABILITY NETWORK (PCSN)

"The money from CFCAS has really provided a wonderful opportunity
and generated useful results—not just data that's going to sit in the bowels
of academia and in journals, but rather data that can be used and ingested
immediately by land managers and provincial and federal governments."

UNIVERSITY OF NORTHERN BRITISH COLUMBIA SCIENTIST BRIAN MENOUNOS,
LEAD INVESTIGATOR WITH THE WESTERN CANADIAN CRYOSPHERIC NETWORK (WC2N)

SELECTED REFERENCES

Introduction: The Search for Answers

Government of Canada. 2007. *Mobilizing Science and Technology to Canada's Advantage*. Ottawa, Ont.: Government of Canada Publications, 2007.

Warren, F.J., and P. Egginton. 2008. "Background Information: Concepts, Overviews and Approaches." *From Impacts to Adaptation: Canada in a Changing Climate 2007*. Eds. Donald S. Lemmen et al. Ottawa, Ont.: Government of Canada Publications, 2008.

Chapter 1: The Storms

CBC News. "Costs of damages from Pangnirtung flood exceed $5M." Tuesday, January 27, 2009. Online at http://www.cbc.ca/canada/north/story/2009/01/27/pang-repairs.html. Retrieved October 21, 2010.

Henstra, D., and G. McBean. *Climate Change and Extreme Weather: Designing Adaptation Policy*. Burnaby, B.C.: Simon Fraser University, 2009.

Honey, K. "1998 saw weathering heights and depths from January's ice storm to December's unusually warm temperatures, Canada marked one of the strangest years on record." *The Globe and Mail*. Toronto, Ont., A.8, December 30, 1998.

Munich Re. "Ambitious climate protection targets are needed – or the cost of climate change will keep rising." November 26, 2009. Online at http://www.munichre.com/en/media_relations/press_releases/2009/2009_11_26_press_release.aspx. Retrieved July 14, 2010.

M.L. Parry, O.F. Canziani, J.P. Palutikof, P.J. van der Linden and C.E. Hanson (eds). *Climate Change 2007: Impacts, Adaptation and Vulnerability, Contribution of Working Group II to the Fourth Assessment Report of the Intergovernmental Panel on Climate Change*. Cambridge, U.K.: Cambridge University Press, 2007.

Chapter 2: The Droughts

Goss Gilroy Inc. *Final Program Evaluation for the Canadian Foundation for Climate and Atmospheric Sciences, Case Studies*. Ottawa, Ont.: Goss Gilroy Inc., 2010.

Hoerling, M., and A. Kumar. 2003. "The Perfect Ocean for Drought." *Science* 299: 691-694.

Michels, A., K.R. Laird, S.E. Wilson, D. Thomson, P.R. Leavitt, R.J. Oglesby, and B.F. Cumming. 2007. "Multi-decadal to millennial-scale shifts in drought conditions on the Canadian prairies over the past six millennia: Implications for future drought." *Global Change Biology* 13: 1,295-1,307.

Mahoney, J. "Drought wounds Prairie farmers." *The Globe and Mail*. Toronto, Ont., A1, July 18, 2002.

Schindler, D.W., and W.F. Donahue. 2006. "An impending water crisis in Canada's western prairie provinces." *Proceedings of the National Academy of Science* 103: 7210–7216.

Solomon, S., D. Qin, M. Manning, Z. Chen, M. Marquis, K.B. Averyt, M. Tignor and H.L. Miller (eds). *Climate Change 2007: The Physical Science Basis, Contribution of Working Group I to the Fourth Assessment Report of the Intergovernmental Panel on Climate Change*. Cambridge, U.K.: Cambridge University Press, 2007.

Wheaton, E., V. Wittrock, S. Kulthreshtha, G. Koshida, C. Grant, A. Chipanshi, B. Bonsal, P. Adkins, G. Bell, G. Brown, A. Howard, and R. Macgregor. *Lessons Learned from the Canadian Drought Years of 2001 and 2002: Synthesis Report*. Saskatoon, Sask.: Saskatchewan Research Council, 2005.

Wheaton, E., S. Kulshreshtha. V. Wittrock, and G. Koshida. 2008. "Dry times: hard lessons from the Canadian drought of 2001 and 2002." *Canadian Geographer / Le Géographe canadien* 52: 241–262.

Chapter 3: The Cities

Hemsworth, W. 2009. "Emergency calls flood in; Narrow escape for women, infants as car becomes engulfed." *The Hamilton Spectator* , Hamilton, Ont., A01, July 27, 2009.

Judek, S., B. Jessiman, D. Stieb and R. Vet. 2004. "Estimated Number of Excess Deaths in Canada Due to Air Pollution." Health Canada, Air Health Effects Division, and Environment Canada, Meteorological Service of Canada. Online at http://www.hc-sc.gc.ca/ahc-asc/media/nr-cp/2005/2005_32bk2_e.html. Retrieved July 11, 2010.

Solomon, S., D. Qin, M. Manning, Z. Chen, M. Marquis, K.B. Averyt, M. Tignor and H.L. Miller (eds). *Climate Change 2007: The Physical Science Basis, Contribution of Working Group I to the Fourth Assessment Report of the Intergovernmental Panel on Climate Change*. Cambridge, U.K.: Cambridge University Press, 2007.

Chapter 4: The North

Arctic Monitoring and Assessment Programme. *Update on Selected Climate Issues of Concern*. Oslo, Norway: Arctic Monitoring and Assessment Programme, 2009.

Blatchford, A. 2010. "Melting permafrost prompts Arctic community housing developments to skip town." The Canadian Press, June 20, 2010. Online at http://www.macleans.ca/article.jsp?content=n3712317. Retrieved October 21, 2010.

England, J. 2010. "Canada needs a polar policy." *Nature* 463: 159

Screen, J. A., and I. Simmonds. 2010. "The central role of diminishing sea ice in recent Arctic temperature amplification." *Nature* 464: 1334-1337.

Solomon, S., D. Qin, M. Manning, Z. Chen, M. Marquis, K.B. Averyt, M. Tignor and H.L. Miller (eds). *Climate Change 2007: The Physical Science Basis, Contribution of Working Group I to the Fourth Assessment Report of the Intergovernmental Panel on Climate Change*. Cambridge, U.K.: Cambridge University Press, 2007.

Tarnocai, C., J. G. Canadell, E. A. G. Schuur, P. Kuhry, G. Mazhitova, and S. Zimov. 2009. "Soil organic carbon pools in the northern circumpolar permafrost region." *Global Biogeochemical Cycles* 23: GB2023, doi:10.1029/2008GB003327. Online at http://www.agu.org/journals/ABS/2009/2008GB003327.shtml. Retrieved October 21, 2010.

Chapter 5: The Ice and Water

Arctic Climate Impact Assessment. *Arctic Climate Impact Assessment Overview Report*. Cambridge, U.K.: Cambridge University Press, 2004.

Bolch, T., B. Menounos, and R. Wheate. 2010. "Landsat-based inventory of glaciers in western Canada, 1985–2005." *Remote Sensing of Environment* 114: 127–137.

Berthier, E., E. Schiefer, G. K. C. Clarke, B. Menounos, and F. Rémy. 2010. "Contribution of Alaskan glaciers to sea-level rise derived from satellite imagery." *Nature Geoscience* 3: 92–95.

Casassa, G., P. Lˊopez, B. Pouyaud, and F. Escobar. 2009. "Detection of changes in glacial run-off in alpine basins: examples from North America, the Alps, central Asia and the Andes." *Hydrological Processes* 23: 31–41.

Moore, R.D., S. W. Fleming, B. Menounos, R. Wheate, A. Fountain, K. Stahl, K. Holm, and M. Jakob. 2009. "Glacier change in western North America: influences on hydrology, geomorphic hazards and water quality." *Hydrological Processes* 23: 42–61.

Smith, S. "BC Hydro predicting dry conditions, aims to change treaty." *Arrow Lakes News*, Nakusp, B.C., June 22, 2010. Online at http://www.bclocalnews.com/ news/96894614.html. Retrieved August 3, 2010.

Chapter 6: The Land

Amiro, B.D., J.B. Todd, B.M. Wotton, K.A. Logan, M.D. Flannigan, B.J. Stocks, J.A. Mason, D.L. Martell, and K.G. Hirsch. 2001. Direct carbon emissions from Canadian forest fires, 1959 to 1999. *Canadian Journal of Forest Research* 31: 512-525.

Bailey, I. "Cost of fighting B.C. forest fires tops $200-million." *The Globe and Mail*. Toronto, Ont., A5, August 19, 2009.

Kurz, W.A., G. Stinson, G.J. Rampley, C.C. Dymond, and E.T. Neilson. 2008. "Risk of natural disturbances makes future contribution of Canada's forests to the global carbon cycle highly uncertain." *Proceedings of the National Academy of Sciences* 105: 1551–1555.

Chapter 7: The Oceans

Ablain, M., A. Cazenave, G. Valladeau, and S. Guinehut. 2009. "A new assessment of the error budget of global mean sea level rate estimated by satellite altimetry over 1993–2008." *Ocean Science* 5: 193–201.

Bender, M.A., T.R. Knutson, R.E. Tuleya, J.J. Sirutis, G.A. Vecchi, S.T. Garner, and I.M. Held. 2010. "Modeled Impact of Anthropogenic Warming on the Frequency of Intense Atlantic Hurricanes." *Science* 327: 454– 458.

Brewster, M. "Hurricane Juan mystifies scientists." *The Globe and Mail*. Toronto, Ont., A5, August 12, 2004.

Fogarty, C. *Hurricane Juan Storm Summary*. Halifax, Nova Scotia: Environment Canada, Canadian Hurricane Centre, 2004. Online at http://www. novaweather.net/Hurricane_Juan_files/Juan_ Summary.pdf. Retrieved October 21, 2010.

Foot, R. "Hurricane Juan hits Halifax in full fury." *Edmonton Journal,* Edmonton, Alta., A1, September 29, 2003.

Nicholls, R., and A. Casenave. 2010. "Sea-level rise and its impact on coastal zones." *Science* 328: 1517-1520.

Solomon, S., D. Qin, M. Manning, Z. Chen, M. Marquis, K.B. Averyt, M. Tignor and H.L. Miller (eds). *Climate Change 2007: The Physical Science Basis, Contribution of Working Group I to the Fourth Assessment Report of the Intergovernmental Panel on Climate Change.* Cambridge, U.K.: Cambridge University Press, 2007.

Chapter 8: The Sky

Mittelstaedt, M. "Statscan finds widespread vitamin D deficiency in Canadians." *The Globe and Mail,* Toronto, Ont. March 24, 2010.

NASA. "Last decade was warmest on record, 2009 one of warmest years, NASA research finds." *ScienceDaily*, January 22, 2010. Online at http://www. sciencedaily.com/releases/2010/01/100121170717. htmScienceDaily. Retrieved October 21, 2010.

Vallis, M. 2010. "Lack of vitamin D costs Canada up to 37,000 lives and $14-billion a year: study." *National Post*, Toronto, Ont. April 7, 2010.

Zeng, G., O. Morgenstern, P. Braesicke, and J. A. Pyle. 2010. "Impact of stratospheric ozone recovery on tropospheric ozone and its budget." *Geophysical Research Letters* 37: doi:10.1029/2010GL042812. Online at http://www.agu.org/pubs/ crossref/2010/2010GL042812.shtml. Retrieved October 21, 2010.

Chapter 9: The World

Hance, J. 2009. "Ban Ki-moon: climate change 'greatest collective challenge we face'." *Mongabay News*. August 10, 2009. Online at http://news. mongabay.com/2009/0810-hance_un_climate.html. Retrieved July 20, 2010.

Conclusion: Questions Remaining

Schiermeier, Q. 2010. "The Real Holes in Climate Science." *Nature* 463: 284-287.

PHOTO CREDITS

Cover, page 1
Ed Darack/
Science Faction/Corbis

page 12
Momatiuk–Eastcott/
Corbis

page 16
Tibor Bognar/Corbis

pages i–ii, 39, v–vi
Josef P. Willems/Corbis

page 13
Jeff McIntosh/CP

page 18
Adrian Wyld/CP

page 8
Steven Kazlowski/
Science Faction/Corbis

page 13
Bob Webber/CP

page 19
Adrien Veczan/CP

page 9
Darrell Gulin/Corbis

page 14
John Ulan/CP

page 20
Paul Souders/Corbis

page 11
Christopher J. Morris/
Corbis

page 14
Mark Taylor/CP

page 24
Moodboard
Photography/Veer

page 27
Christopher Morris/
Corbis

page 33
Paul Darrow/
Reuters/Corbis

page 37
Mario Beauregard/CP

page 28
Michael Christopher
Brown/Corbis

page 34
Alloy Photography/Veer

page 40
Igor Korionov/Veer

page 29
Darryl Dyck/CP

page 34
Richard Cummins/
Corbis

page 41
Heino Kalis/
Reuters/Corbis

page 30
EPPIC/Veer

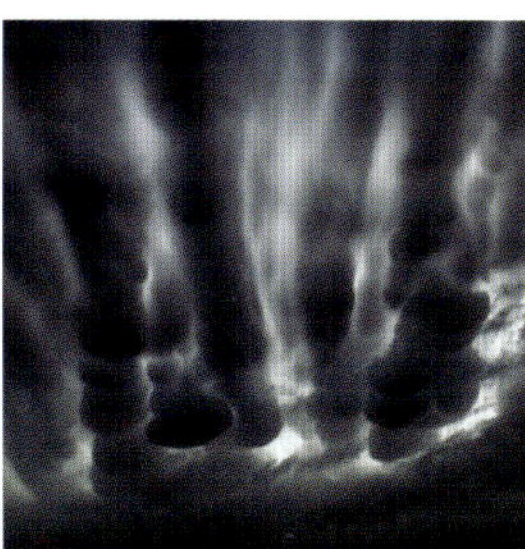

page 36
Corey Hochachka/
Design Pics/Corbis

page 52
Philippe Renault/
Hemis/Corbis

page 32
Barrett & MacKay/
All Canada Photos/
Corbis

page 37
Douglas Kirkland/
Corbis

CFCAS also thanks its staff and funded researchers for other photos that appear in this book.

INDEX